Elka Costa Santos Nascimento
Viviane Farias Silva
Kalyne Sonale Arruda Brito

Seedlings of sunflower cultivars produced with waste

Elka Costa Santos Nascimento
Viviane Farias Silva
Kalyne Sonale Arruda Brito

Seedlings of sunflower cultivars produced with waste

Producing flowers with waste, economy and sustainability in agriculture

ScienciaScripts

Imprint

Any brand names and product names mentioned in this book are subject to trademark, brand or patent protection and are trademarks or registered trademarks of their respective holders. The use of brand names, product names, common names, trade names, product descriptions etc. even without a particular marking in this work is in no way to be construed to mean that such names may be regarded as unrestricted in respect of trademark and brand protection legislation and could thus be used by anyone.

Cover image: www.ingimage.com

This book is a translation from the original published under ISBN 978-3-330-75368-6.

Publisher:
Sciencia Scripts
is a trademark of
Dodo Books Indian Ocean Ltd. and OmniScriptum S.R.L publishing group

120 High Road, East Finchley, London, N2 9ED, United Kingdom
Str. Armeneasca 28/1, office 1, Chisinau MD-2012, Republic of Moldova, Europe
Printed at: see last page
ISBN: 978-620-8-27222-7

SUMMARY

1 INTRODUCTION

Sunflower cultivation is increasing significantly every year, mainly due to its unique characteristics of hardiness, drought resistance, beauty, oil content and quality (SOUZA et al., 2010). In northeastern Brazil, efforts have been made to establish sunflower cultivation practices that make it viable to exploit under rational and economical techniques, mainly because it is a plant with high nutritional requirements (SOUZA *et al.*, 2010).

Sunflower can be used in a variety of ways, including for medicinal purposes, green manure, crop rotation, beekeeping, animal feed, oil production, human food and biodiesel production (SILVA *et al*, 2011).

For a material to be used as a substrate for seedlings, according to Severino *et al.* (2006) it must have appropriate chemical and physical characteristics, be available nearby in sufficient quantity and be low cost. The addition of organic matter to the soil helps to improve its physical structure, water absorption capacity and the supply of nutrients to plants, making it possible to increase production and improve food quality (MALHEIROS and PAULA JÙNIOR, 1997).

Spadotto & Ribeiro (2006) state that the agro-industry is one of the main segments of the Brazilian economy, with importance in both domestic supply and exports. In this context, the waste generated in these activities has the potential to impact on the environment if not treated properly. The environmental impacts associated with these residues stem from their high generation in quantitative terms and slow degradability in some cases, and in others, from the generation of by-products that can be toxic, cumulative or difficult to degrade (IPEA, 2012).

Once generated, waste needs to be properly disposed of because, as well as creating potential environmental problems, waste represents a loss of raw materials and energy, requiring significant investment in treatments to control pollution. The food industry produces several residues with a high reuse value (PELIZER *et al.*, 2007).

This research was carried out in order to determine the ideal substrate composition using agro-industrial waste for the production of sunflower cultivars.

2. LITERATURE REVIEW

2.1. Wastewater for irrigation

The use of wastewater for agriculture has some advantages, as stated by Van Der Hoek *et al.* (2002), in the conservation of available water and the possibility of contributing and recycling nutrients, resulting in the preservation of the environment. The practice of reusing this water for irrigation has been an excellent measure to alleviate the problem of water scarcity in the Brazilian semi-arid region (SOUSA and LEITE, 2003).

Oliveira *et al.* (2012) studied the effect of treated wastewater on the production of Cambuci pepper seedlings and okra, an economically viable alternative for use in irrigation, with a significant effect on the variables analyzed. Research on cowpeas irrigated with wastewater, Rebouças *et al.*, (2010) found a positive effect on total phytomass in plants irrigated only with effluent, with an increase in total dry matter production of around 117.0%, and concluded that the nitrogen in the treated wastewater used for irrigation met the water and nutritional needs. Satisfactory results were obtained by Augusto *et al.* (2003) studying the production of *Croton floribundus Spreng.* (Capixingui) and *Copaifera langsdorfii. Desf.* (copaiba) in a sub-irrigation system with treated wastewater.

Other studies have also been carried out with the aim of verifying the results of using wastewater in irrigation, such as Augusto *et al.*, (2007) who compared the production of seedlings irrigated with wastewater and with fertilizer application and found no apparent mortality, deficiency or toxicity, as well as Alves *et al.*, (2009) analyzing cotton irrigated with wastewater, affirmed that wastewater did not affect the development of cotton plants. Comparing the annual yield of some crops, Shende (1985) noted that crops fertigated with wastewater had higher yields than crops irrigated with clean water and fertilized with chemical fertilizers.

Some of the changes in soil irrigated with wastewater mentioned by Fonseca *et al.* (2007) are the effects on total carbon and nitrogen, microbial activity and N-minerals, exchangeable calcium and magnesium, salinity, sodicity and clay dispersion. The application of this inferior quality water for irrigation, when carried out indiscriminately, without agronomic and environmental criteria, can cause problems of contamination of the soil, surface water and groundwater and toxicity to plants; however, if well planned, this application will bring benefits, according to Erthal *et al.,* (2010) such as: nutrients and water for the plants, a reduction in fertilizers and their polluting potential. According to Cabral *et al.* (2011), information on the use of secondary waters in agricultural areas, environmental impacts, forms of management, recovery of degraded areas, alternative treatment of these waters, is indispensable.

Otenio (2015) reports that the agro-industry needs a large amount of water in its process, resulting in a certain amount of wastewater and this effluent can come from the sanitization of facilities with a high rate of Chemical Oxygen Demand (COD) and Biological Oxygen Demand (BOD), for this reason there is interest in management and treatment solutions, either for reuse or for disposal in the environment.

There is specific legislation for wastewater, such as COMANA Resolution 430/2011 (BRAZIL, 2011), which only establishes quality criteria for discharge into surface water bodies, but does not establish quality criteria for reuse. However, the World Health Organization (WHO) classifies the types of reuse into different modalities, according to their uses and purposes (WHO, 1973). Resolution 54 of the National Council of Water Resources (CNRH) established modalities, guidelines and general criteria for the practice of secondary water reuse, with the aim of expanding the use of low quality water (BRASIL, 2006).

Due to the scarcity of water resources, alternatives and technologies have been developed to treat wastewater for reuse in irrigated agriculture (AQUINO NETO *et al.,* 2011).

2.2. Agro-industrial waste

Agribusiness in Brazil is gaining prominence every year, but this growth also generates waste that must be included in the company's planning, as Rosa *et al.* (2011) state that Brazilian agribusiness puts the country in a prominent position in the development process. Since the industrial revolution, which helped to carry out work in the fields, making it possible to increase the area under cultivation and consequently production, with technological investments in the sector, the quality and quantity of grain has doubled, as Gasques and Bastos (2003) show.

This is due to the great potential for agricultural production, forestry and waste generated by the agro-industry, such as rice husks, wheat husks, sugar cane bagasse, among others. The importance of the agro-industry in the gross domestic product (GDP), in exports and in controlling inflation are examples of the value of this sector to the performance of the Brazilian economy, according to Pessôa, (2009).

Agriculture and agro-industry play a fundamental role in the history, quality of life and economy of a country, but there is concern about the waste that is generated, which has a negative impact on the environment. In this way, efficient treatment of these end products is important for the conservation of the environment and the sustainability of agriculture. Some authors, such as Malheiros and Paula Júnior (1997), describe that the environmental impacts associated with these residues are due to their high generation in quantitative terms and their slow degradability, which can be toxic, cumulative or difficult to degrade. In this case, the use of the 3R's, reduce, recycle or reuse the residues generated in order to recover material and energy, preserving natural resources and avoiding environmental degradation.

Rural waste is considered to be any type of waste produced in rural areas as a result of agricultural management, and this quantification is based on harvest indices, which is the ratio between the total amount of biomass produced per hectare and the amount of product according to Residuos no Brasil (2016), planted from a given crop and the amount of product.

It is estimated that 250 million tons of agro-industrial waste are produced every year and that reusing it would help to reduce environmental and energy problems, with the possibility of producing products of great importance in other sectors, such as the pharmaceutical and food industries (ROBERTO *et al.*, 1995; SOUZA and SANTOS, 2002), as well as fertilizers. In 2015, around 21.0 million tons of fertilizers were sold in Brazil, a large percentage of which were used to meet the demands of soybean, sugar cane, coffee, corn and other crops, according to the National Association for the Dissemination of Fertilizers (ANDA, 2016). In 2016, this figure already stood at 22.3 million tons of imported fertilizers (ANDA, 2016). We can see the country's dependence on an essential agricultural implement and the price of the product, so when applying alternatives as a source of fertilizer, agro-industrial waste becomes an important factor to be taken into account, making cultivation less costly and independent, as well as sustainable.

Pereira (2006) states that around 350 million tons of agricultural waste are produced annually in Brazil, with sugarcane waste generating the most. In the 2007/2008 harvest, around 500 million tons of sugarcane were produced in Brazil (UNICA, 2008) and in the 2015/2016 harvest there were 666.8 million tons, with the largest production in the Center-South region with 617.7 million tons and the North-Northeast region with 49.1 million tons (UNICA, According to Burgi (1995), every ton of sugarcane crushed in the industry produces 700 liters of sugarcane juice and 300 kg of bagasse. It is therefore clear that the growing demand for production directly increases the production of waste, making it important to dispose of it properly.

As Aguiar and Ferraz (2011) state, agro-industrial waste has important characteristics, due to the way in which raw materials with greater added value are processed. Agricultural waste is made up of around 60% cellulose, 30% hemicellulose and 30% lignin. Sugarcane bagasse and rice husks, for example, are made up of approximately 40% cellulose, 35% hemicellulose and 35% lignin (OLIVEIRA *et al.*, 2003). According to Carvalho (1992), these agro-industrial residues and residues from the processing of plant products can be used to feed ruminants, and are an option in times of extreme weather conditions, such as drought and winter. The sugar-energy units are milling to produce sugar and ethanol and generating energy from burning the bagasse, with the waste generated being a source of energy (HOLLAND A, 2013).

There are a number of advantages to using agro-industrial residues and for sugar cane bagasse we can mention, according to Pellegrini (2002) :a) the sugarcane plantation is organized and the costs of producing bagasse are charged to the noble product, sugar and/or alcohol; b) the transport system between the field and the industry is organized and linked to the same noble product; c) bagasse production is large and concentrated and the product is released semi-processed after the sugarcane has been crushed. The Union of the Sugar Cane Industry (UNICA, 2012) states that one ton of sugar cane produces an average of approximately 280 kg of bagasse and 234 kg of straw and tips, so with bagasse alone it would be possible to generate around 85.6 kWh of energy for export, using a 65 bar boiler producing 342.4 kWh for export with one ton of bagasse, while one ton of straw generates 500 kWh due to the lower humidity, as reported by Nogueira and Garcia (2013).

Leao *et al.* (2011) define tailings as an inadequate fraction of waste and its final disposal must take into account its inactivation, neutralization, decontamination or detoxification. There are various options for this final disposal, such as recycling, landfill, composting and other forms. These authors state that the relationship between the final disposal options would result in a product with greater added value and an additive in the plant production chain, generating a material with economic and environmental value. In this way, the production and use of alternative materials from agro-industrial waste becomes feasible, simply because it reduces the costs for the final disposal of waste, as well as producing a product with various uses.

Some examples found in the literature are alternatives for agricultural waste, such as a substrate for growing orchids based on coffee husks (ASSIS, 2011), reducing production costs and environmental impacts, and by-products derived from fruit in cattle feed with appreciable results (AZEVEDO, 2011). The use of waste in agriculture, according to Pires and Mattiazzo (2008), requires planning, based on some legal, environmental and economic aspects, verifying that this use should be economically viable and environmentally sustainable. In research carried out by Sousa *et al.* (2010) on the evaluation of lignocellulosic residues such as banana peels, stalks, pseudostems and banana leaves as a fermentation substrate in the methanization process, it was found that they have the potential to be used as a substrate to produce biogas, with great technical and economic viability.

Some fibrous fraction residues can be a source of energy for ruminants, according to SOUZA *et al.,* (2005) lignocellulosic residues are one of these sources, due to the breaking of the bond between lignin and cellulose and hemicellulose, increasing the availability of energy within the rumen. Various agro-industrial residues are applied in agriculture, such as ornamental sunflower with different doses of chicken litter, Andrade *et al.* (2010), and the interaction of organic substrate and wastewater, verified by Souza *et al.* (2010), analyzing quantities of earthworm humus and domestic effluent.

2.3. Sunflower cultivation

Belonging to the tribe Heliantheae, family Asteraceae and order Asterales, the cultivated sunflower (*Helianthus annuus* L.) has a chromosome number of 2n=34. Sunflower comes from the Greek and means sun flower. The plant's characteristic is to move in line with the movement of the sun, known as heliotropism, as stated by the authors Nunes *et al.* (2013).

Some researchers claim that the sunflower originated in Peru, but there are archaeological studies that affirm the use of this crop by North American Indians, with at least one indication of cultivation in the states of Arizona and New Mexico, around 3000 years BC (SELMECZI-KOVACS *et al.,* 1975). There are also stories about the Spanish colonizers who called the sunflower the flower of Peru (VRÂNCEANU, 1977) and speculation that the sunflower was domesticated before corn, as reported by Putt (1997).

Lentz et al. (2001) describe that the oldest sunflowers were found at an archaeological site in Mexico and that studies have found that the sunflowers used today came from very restricted genetics, which may suggest that they came from only one domestication, the most accepted hypothesis being that the sunflower arose from the wild sunflower which was domesticated and used to feed the natives. Putt (1977) describes that sunflower seeds were ground into flour to make bread, and that the seeds were also used to produce purple paint for decoration, body paint and hair for beautification in religious rituals, and that the roots and capitula were used for medicinal purposes. This shows the importance of the sunflower in the development of the region's populations, as all parts of the plant were used.

Sunflower has several uses, which have influenced the increase in the area cultivated in Brazil (CONAB, 2011), as well as its characteristics of adaptation to different climatic conditions, drought tolerance, among others (EMBRAPA, 2005), as well as its adaptability to various edaphic characteristics and its yield, which is insignificantly influenced by altitude and altitude (CASTRO *et al.*, 1997). According to Ungaro (2000) it is an annual plant with a pivoting root system.

The importance of this crop in the economic and industrial environment is clear (Nunes *et al.*, 2013). The production of sunflower oil for human consumption can prevent heart disease, according to Leite (2005). With concern for the environment, researchers have looked to the sunflower as an alternative for producing a fuel with lower pollution levels, which would reduce the use of fossil fuels and generate a sustainable fuel, which has been achieved with sunflower biodiesel (GOLDEMBERG, 2008; NEVES, 2011).

The production of biodiesel using sunflower as a raw material must be within the established standards (NEVES, 2011) because the type of fatty acid and its concentration in triglycerides are directly influenced by the raw material, just as the types of oils derived from vegetables are influenced by the environment and genetics that have developed (PIMENTEL, 2007).

Sunflower cultivation is important due to its multiple uses such as: ornamental flower, confectionery sunflower as a substitute for almonds *in* general, *fresh* grains and bran for feeding poultry, pigs and cattle, fodder, silage, consumed in fresh, toasted, salted and packaged human food, however the development of the crop is currently linked to oil extraction (GAZOLLA *et al.*, 2012). These authors also state that the plant's root system recycles nutrients in the soil, and that the stems are used to make material for acoustic insulation, green manure, and to produce honey from the flowers, among other things.

The cultivation of sunflower with lower quality water is being researched due to the greater economic potential observed by some authors, Oliveira *et al.*, 2010; Nobre *et al.*, 2010; Campos *et al.*, 2010; Travassos *et al.*, 2011; Santos Jûnior *et al.*, 2011, for the great diversity of use of all parts of the plant and various purposes. For floriculture, due to the plant's characteristics, according to Huang (1995) it has become an option for ornamentation. Sunflowers produced for ornamental purposes with around six stems are sold on average for around R$15.00, depending on the time of year, as Curti (2010) explains.

The National Supply Company (CONAB) began recording cultivated areas/region from 1997 onwards, noting that it was only in the Central West, Southeast and South regions that sunflower production began, and there has been a growing trend in sunflower harvests, as shown in Table 1.

Table 1 shows that the North/Northeast region does not have a significant production, with records only in the Northeast region in the 2009/2010 and 2011/2012 harvests. The regions that have contributed the most since the 1997/1998 harvest is the Midwest, with the states of Mato Grosso and Goiás as the largest producers.

Table 1. Crop records and sunflower growing regions in Brazil.

REGION/UF	1997/98	1999/00	2003/04	2005/06	2009/10	2011/12	2013/14	2014/15	2015/2016	2016/17 Forecast
NORTH	-	-	-	-	-	-	-	-	-	-
NORTH EAST	-	-	-	-	1,4	0,2	-	-	-	-
WEST CENTER	9,6	53,4	45,8	43,5	55,8	66,0	131,1	94,2	40,9	40,9
MT	2,7	3,2	9,3	17,3	40,60	47,10	126,20	86,40	25,6	25,6
MS	1,8	16,1	13,1	18,5	3,8	5,0	0,7	0,4	1,3	1,3
GO	5,1	34,1	23,1	7,4	11,4	13,9	4,2	7,4	14,0	14,0
DF	-	-	0,3	0,3	-	-	-	-	-	-
SOUTHEAST	0,7	1,5	2,0	2,2	-	4,3	11,3	14,0	7,0	7,0
MG	-	-	-	-	-	4,3	11,3	14,0	7,0	7,0
ES	-	-	-	-	-	-	-	-	-	-
RJ	-	-	-	-	-	-	-	-	-	-
SP	0,7	1,5	2,0	2,2	-	-	-	-	-	-
SOUTH	2,1	3,1	7,3	21,2	13,8	4,0	3,3	3,3	3,3	3,0
PR	1,2	1,5	0,3	1,2	0,7	0,7	-	-	-	-
SC	-	-	-	-	-	-	-	-	-	-
RS	0,9	1,6	7,0	20,0	13,1	3,3	3,3	3,3	3,3	3,0
NORTH/NORTHEAST	-	-	-	-	1,4	0,2	-	-	-	-
SOUTH CENTRAL	12,4	58,0	55,1	66,9	69,6	74,3	145,7	111,5	51,2	50,9

BRAZIL	12,4	58,0	55,1	66,9	71,0	74,5	145,7	111,5	51,2	50,9

Source: CONAB, 2016.

3. BIBLIOGRAPHICAL REFERENCES

AGUIAR, A., FERRAZ, A. (2011). Review: Mechanisms involved in the biodegradation of lignocellulosic materials and related technological applications. **Quim. Nova**, 34:1729-1738.

ALVES, W. W. A. et al. Leaf area of cotton irrigated with wastewater fertilized with nitrogen and phosphorus. Revista Verde de Agroecologia e Desenvolvimento Sustentàvel, Mossoró-RN, v. 4, n. 1, p. 41-46, 2009.

ANDA, National Association for the Diffusion of Fertilizers, 2016. http://anda.org.br/index.php?mpg=03.00.00&ver=por

ANDRADE, L. O. et al. Production of colored sunflower as a function of fertilization with chicken litter and irrigation with treated domestic effluent. In: WINOTEC - INTERNATIONAL WORKSHOP ON TECHNOLOGICAL INNOVATIONS IN Irrigation, Fortaleza, 2010, Anais... Fortaleza: INOVAGRI. 2010.

AQUINo NETo, s.de; MAGRI, T.c.; sILVA, G.M.da; ANDRADE, A.R.de. Treatment of dye residues by electroflocculation: an experiment for undergraduate chemistry courses. **Quimica Nova** [online]. 2011, v.34, n.8, p.1468-1471.

ASSIS, Adriane Marinho de et al . Orchid cultivation in coffee husk-based substrates. **Bragantia**, campinas, v.70, n.3, 2011.

AUGUSTo, D. c. c. et al. Use of domestic sewage treated through a biological system in the production of seedlings of Croton floribundus Spreng. (Capixingui) and Copaifera langsdorfii. Desf. (Copaiba). **Revista Arvore**, v.27, n.3, p.335-342, 2003.

AUGUSTO, D. C. C., et al. Use of wastewater from the biological treatment of domestic sewage in the production of Eucalyptus grandis hill seedlings. **Revista Arvore**, Viçosa-MG, v.31, n.4, p.745-751, 2007.

AZEVEDO, J.A.G. et al. Intake, total digestibility, microbial protein production and nitrogen balance in diets with fruit by-products for ruminants. **Revista Brasileira de Zootecnia**, Viçosa, v.40, n.5, May 2011.

BRAZIL. Ministry of the Environment. National Water Resources Council. Resolution No. 54 of November 28, 2005. Establishes modalities, guidelines and general criteria for the practice of direct non-potable reuse and other measures. Official Journal of the Union, Brasilia, 09/03/2006.

National Environment Council. Resolution No. 430, of May 13, 2011. Provides for the conditions and standards for effluent discharge, complements and amends Resolution No. 357, of March 17, 2005, of the National Environment Council - CONAMA. Brasilia, 2011.

BURGI, R. Utilization of crop and processing residues in cattle feed. In: Proceedings of the 6th FEALQ symposium on cattle nutrition. Piracicaba-SP: FEALQ, 1995. p. 153-169.

CABRAL, J.R.; FREITAS, P.S.L.; REZENDE, R. et al. Impact of pig wastewater on soil and elephant grass production. **R. Bras. Eng. Agric. Ambiental**, v.15, n.8, p.823-831, 2011.

CAMPOS, M. S.; OLIVEIRA, F. A.; OLIVEIRA, F. R. A.; SILVA, R. C. P.; CANDIDO, W. S. Effect of salinity and nitrogen sources on sunflower dry matter. **Revista Verde de Agroecologia e Desenvolvimento Sustentâvel**, v.5, p165-171, 2010.

CARVALHO, F.C. Availability of agro-industrial residues and the processing of agricultural products. Informaçoes Econômicas, SP, v.22, n.12, dec. 1992.

CASTRO,C.; CASTIGLIONI,V.B.R.; BALLA, A.; LEITE, R.M.V.B.C.; KARAM, D., MELLO, H.C.; GUEDES, L.C.A.; FARIAS, J.R.B. Adubaçâo. In: The sunflower crop. Londrina-PR. Editora EMBRAPA,1997, CAP.7, P.17-19.

CONAB - National Supply Company. Available at . Accessed on: May 26, 2011.

CURTI, G. L. Characterization of ornamental sunflower cultivars sown at different times in western Santa Catarina. Pato Branco: Federal University of Technology of Paranà. 2010. 76p. Master's dissertation

EMBRAPA. National Soybean Research Center. Subsidies for the elaboration of the

National Sunflower Research Program. Londrina, 2005. 17p.

ERTHAL, V.J.T.; FERREIRA, P.A.; MATOS, A.T. et al. Physical and chemical alterations of an Argissolo by the application of bovine wastewater. Revista Brasileira de Engenharia Agricola e Ambiental, v.14, n.5, p.467-477, 2010.

FONSECA, A. F. ; HERPIN, U; PAULA, A. M. et al. Agricultural use of treated sewage effluents: Agronomic and environmental implications and perspectives for Brazil. Scientia Agricola, v. 64, n.2, p.194-209, 2007.

GASQUES, J. G.; BASTOS, E. T. (March/2003). Agricultural Growth. Conjuncture Bulletin, No. 60.

GAZZOLA, A.; FERREIRA JR, C.T.G.; CUNHA, D.A.; BORTOLINI, E.; PAIAO, G.D.; PRIMIANO, I.V.; PESTANA, J.; DANDREA, M.S.C.; OLIVEIRA, M.S. The sunflower crop. Department of Plant Production - University of Sao Paulo, Picacicaba-SP, 2012.

GOLDEMBERG, J. Bioenergy in the state of São Paulo : current situation, prospects, barriers and proposals / José Goldemberg, Francisco E. B. Nigro, Suani T. Coelho - Sao Paulo : Official Press of the State of Sao Paulo, 2008. 152p.

HUANG, M.C. New ornamental crops in Asia. **Acta Horticulturae**, v.397, p.49, 1995.

LEÂO, A.L.; CHERIAN, B.M.; SOUZA, S.F.; THOMAS,S. The production of alternative materials from agro-industrial waste is a viable alternative for reducing the costs of waste disposal. **Revista Citricultural Atual**, n.80, 2011.

LEITE, R.M.V.B.C.; Brighenti, A.M.; Castro, C. Eds. Sunflower in Brazil, Londrina: Embrapa Soja, 2005.

LENTZ, D.; POHL, M.E.D.; POPE, K.O.; WYATT, A.R. Prehistoric sunflower (*Helianthus annuus* L.) domestication in Mexico. **Economy Botany**, New York,v.55, n.3,p.370-376, 2001.

MALHEIROS, S. μ. P.; PAULA JÛNIOR, D. R. Use of the composting process with agro-industrial waste. In: CONGRESSO BRASILEIRO DE CIÊNCIA DO

SOLO, 26, 1997.

NEVES, M. F. (Coordinator). Agribusiness and sustainable development: An agenda for world leadership in food and bioenergy production - 1. ed. - 4.reimpr. - Sao Paulo: Atlas, 2011.

NOBRE, R. G.; GHEYI, H. R.; CORREIA, K. G.; SOARES, F. A. L.; ANDRADE, L. O. Sunflower growth and flowering under salt stress and nitrogen fertilization. **Revista Ciência Agronòmica**, v. 41, p.358-365, 2010.

NOGUEIRA, M.A.F.S.; GARCIA, M.S. Waste management in the sugar-energy industrial sector: case study of a mill in the municipality of Rio Brilhante, Mato Grosso do Sul. **Revista Eletronica em Gestâo, Educaçâo e Tecnologia Ambiental** - REGET, v. 17 n. 17, 2013, p. 3275 - 3283

NUNES, B.T.; PILON, A.; FLUMINHAN, A. Evaluation of the agronomic performance of sunflower genotypes (*Helianthus annuus* L.) grown in the Oeste Paulista region and analysis of oil, dry matter and crude protein yields. **Electronic journal Fórum Ambiental da Alta Paulista**, v.9, n.1, p.150-165, 2013.

OLIVEIRA, F. A.; OLIVEIRA, F. R. A.; CAMPOS, M. S.; OLIVEIRA, M. K. T.; MEDEIROS, J. F.; SILVA, O. M. P. Interaction between salinity and nitrogen sources in the initial development of the sunflower crop. **Revista Brasileira de Ciências Agrârias**, v.5, p.479-484, 2010.

OLIVEIRA, J.F. et al. Effect of treated domestic sewage wastewater on the production of Cambuci pepper and okra seedlings. **Enciclopédia Biosfera**, v.8,n.14,p.443-452, 2012.

OLIVEIRA, R. M.; CARVALHO, E. P.; SCHUWAN, R. F. Production of extracellular hydrolytic enzymes by Fusarium in a simple batch system. **Revista Interaçâo**, Juiz de Fora, n.2, 2000.

PEREIRA, R. E. Avaliaçao do Potencial Nacional de Geraçao de Residuos Agricolas para a Producao de Etanol. 2006. Dissertation (**Master of Science**) - School of Chemistry, Federal University of Rio de Janeiro - UFRJ, Rio de Janeiro.

PESSÔA, A. Agricultura. Brasilia: MRE, 2009.

PIMENTEL, M. S. Special Biodiesel - Wider Margins. In: Panorama Rural. pp. 55 - 69, 2007.

PIRES, A.M.M.; MATTIAZZO, M.E. Avaliação da Viabilidade do Uso de Residuos na Agricultura. Technical Circular 19. EMBRAPA Jaguariûna, Nov 2008. 9p. Rio de Janeiro, v. 14, n.4, 2009.

PUTT, E.D. Early history of sunflower. *In*: SCHNEITER,A.A.(Ed). Sunflower technology and production. Madison: American Society of Agronomy Press, p.1-19, 1997.

REBOUÇAS, J. R. L. et al. Growth of cowpea irrigated with treated domestic sewage wastewater. **Revista Caatinga**, Mossoró, v.23, n. 1, p. 97102, 2010.

WASTE IN BRAZIL.

http://infoener.iee.usp.br/scripts/biomassa/br_residuos.aspDisponivel at: . Accessed on September 18, 2016

ROBERTO, I. C.; FELIPE, M. G. A.; MANCILHA, I. M.; VITOLO, M.; SATO, S.; SILVA, S. S. Xylitol production by Candida guilliermondii as an approach for the utilization of agroindustrial residues. **Bioresource Technology**, Essex, n.51, p.255-257, 1995.

ROSA, M.F.; SOUZA FILHO, M.S.M.; FIGUEIREDO, M.C.B.; MORAIS, J.P.S.; SANTAELLA, S.T.; LEITÂO, R.C. Valorizaçao de residuos da agroindustria. In: II Simposio Internacional sobre Gerenciamento de Residuos Agropecuàrios e Agroindustriais, Foz do Iguaçu-PR, 2011.

Santos Jùnior, J. A.; Gheyi, H. R.; Guedes Filho, D. H.; Dias, N. da S.; Soares, F. A. L. Sunflower cultivation in a hydroponic system under different salinity levels. **Revista Ciência Agronòmica**, v.42, p.842-849, 2011.

SELMECZI-KOVACS, A. Akklimatisation und verbreitung der sonnenblume in Europa. **Acta Ethnographia Academiae Hungaricae**, v.24, n.1-2, p.47-88, 1975.

Shende, G. B. Status of wastewater treatment and agricultural reuse with special reference to Indian experience and research and development needs. In: FAO Regional Seminar on the Treatment and Use of Sewage Irrigation. Rome: FAO, 1985, p. 157-182.

SOUSA, J. T. de; Leite, V. D. Tratamento e utilizaçao de esgotos domésticos na agricultura. Campina Grande: EDUEP, 2003. 135p.

SOUZA, O. et al. Importância da utilizaçao de residuos agropecuárias. Uberaba, MG: O Berro, 2005.

SOUZA, O.; FEDERIZZI, M.; COELHO, B.; WAGNER, T. M.;WISBECK, E. Biodegradation of lignocellulosic residues generated in banana growing and their valorization for biogas production. Revista Brasileira de Engenharia Agricola e Ambiental v.14, n.4, p.438-443, 2010.

SOUZA, O.; SANTOS, I. E. Importance of agricultural waste in animal feed. **Livestock bulletin**. 2002. Available at: . Accessed on: October 11, 2003

SOUZA, R. M. et al. Use of wastewater and organic fertilizer in sunflower cultivation. **Revista Caatinga**, Mossoró, v. 23, n. 2, p. 125-133, 2010.

Travassos, K. D.; Soares, F. A. L.; Gheyi, H. R.; Silva, D. R. S.; Nascimento, A. K. S.; Dias, N. da S. Water yield of sunflower irrigated with brackish water. **Revista Brasileira de Engenharia Agricola e Ambiental**, v.15, p.371-376, 2011.

UNGARO, M.R.G. The sunflower in Brazil. O Agronômico, Campinas, v.34, p.4362, 1982.

UNICA. 2007/08 harvest statistics.

UNICA. 2015/2016 harvest statistics.

VAN DER HOEK, W. et al. Urban wastewater: a valuable resource for agriculture. A case study from Horoonabad, Pakistan. Colombo, Sri Lanka: **International Water Management Institute**, 2002. 29 p. (Research Report, 63).

VRANCEANU, A.V. The sunflower. Madrid: Ediciones Mundi-Prensa, 1997.375p.

WHO. Reuse of effluents: methods of wastewater treatment and health safeguards. Report of a WHO Meeting of Experts. Geneva, World Health Organization (Technical Report Series No. 517), 1973.

CHAPTER 1

DIFFERENT SUBSTRATES FOR GROWING SUNFLOWER IRRIGATED WITH WATER OF VARYING QUALITY

ABSTRACT: Considering the growth of the flower agribusiness in Brazil and due to the scarcity of studies on the use of wastewater and the use of agricultural substrates, the objective was to evaluate the influence of wastewater and combinations of agricultural substrates on the growth and phytomass of ornamental sunflower seedlings (*Helianthus annuus* L.). The work was carried out in a greenhouse at the Federal University of Campina Grande - PB; the following treatments were studied: S1 - 100% commercial substrate, S2 - 100% soil, S3 - 100% coconut fiber, S4 - 50% soil + 50% coconut fiber, S5 - 50% commercial substrate + 50% coconut fiber and S6 - 50% commercial substrate + 50% soil combined with two types of water - supply water and wastewater. The experimental design was in randomized blocks, in a 6 x 2 factorial scheme, with 3 replications. The best results were obtained for sunflower seedlings grown with the combination of commercial substrate + soil and soil + coconut fiber. Wastewater promoted considerable increases in sunflower growth and phytomass.

Keywords: coconut fiber; *Helianthus annuus L.;* commercial substrate

INTRODUCTION

The sunflower originated in southwestern Mexico, where it grows naturally. The species was introduced to Europe in the 14th century as a cultivated plant and reintroduced to America from Europe in the 19th century (SALUNKHE and DESAI, 1986). Currently, the technology for large-scale sunflower production and the appropriate genetic material allow for high grain and oil yields. The sunflower crop has been increasing its production significantly every year, mainly due to its peculiar characteristics of hardiness, resistance to drought, beauty, oil content and quality (SOUZA *et al.*, 2010). In addition to these characteristics, ornamental sunflower has been attracting various lines of research with the aim of improving its agronomy, and is responsible for increasing the flower agribusiness in Brazil (FIGUEIREDO *et al.*,

2008).

Most of Brazil is suitable for its cultivation, making it an option for crop rotation, with advantages over other plants due to its resistance to drought and low temperatures (UNGARO, 2000). In northeastern Brazil, efforts have been made to establish sunflower cultivation practices that make it viable to grow it using rational and economical techniques, mainly because it is a plant with high nutritional requirements (SOUZA *et al.*, 2010). In the state of Paraiba, the municipalities of Alhandra and Conde stand out as producers of tropical ornamental plants. In agreste Paraibano, the municipality of Lagoa Seca is a producer of various flowers (OLIVEIRA & BRAINER, 2007), which confirms the growing expansion of the activity and an increase in the importance of the regional economy.

According to Nascimento and Heller (2005), two thirds of the world's population will be affected by water shortages in the coming decades. However, there are a number of alternatives that could alleviate this problem, including water reuse, control of physical losses in supply systems, rainwater harvesting techniques and the adoption of water-saving procedures.

The adoption of the planned reuse of domestic wastewater in irrigated agriculture, for example, has been considered a prodigious means of alleviating the problem of water scarcity in the semi-arid region, especially in areas surrounding cities (SOUSA *et al.*, 2006), reducing environmental impacts, since the nutrients present in reused water (such as nitrogen), when released into aquifers, can cause severe environmental contamination.

According to Rosa *et al.* (2002), agricultural crops have recently been grown in a substrate based on dried coconut shells, a nutritionally rich material from the agricultural industry which aims to offer an alternative for seedling producers to replace soil, mainly by minimizing the environmental impact caused by the solid agro-industrial waste generated. Also according to Carrijo *et al.* (2002), its good physical properties, its non-reaction with nutrients from fertilization and its long durability without altering its physical characteristics, as well as the abundance of renewable raw materials and the low cost to the producer, make green coconut fibre a substrate that can hardly be surpassed by any other type of substrate, mineral or organic, for growing vegetables and flowers without soil. Commercial substrates, on the other hand, contain elements such as vermiculite, pine bark and carbonized rice husk, mostly in a 1:1:1 ratio.

Consequently, some segments of society have been engaged in developing research aimed at making economic use of waste, which often has the potential for agricultural use, especially as substrates in the production of seedlings (SOUZA, 2001), reducing production costs and minimizing negative environmental impacts.

Considering the growth of the flower agribusiness in Brazil and the scarcity of studies on the use of wastewater and agricultural substrates, the objective was to evaluate the influence of different water qualities and different agricultural substrates on the growth and phytomass of ornamental sunflower seedlings (*Helianthus annuus* L.).

MATERIAL AND METHODS

The experiment was carried out in a greenhouse belonging to the Academic Unit of Agricultural Engineering (UAEAg) of the Federal University of Campina Grande - UFCG, located in the municipality of Campina Grande, state of Paraiba - PB, at the geographical coordinates 7°15'18" south latitude and 35°52'28" west longitude, at an altitude of 550 m.

The seeds of the ornamental sunflower cultivar Sol Noturno and, subsequently, the sunflower seedlings were sown and cultivated in tubes (black plastic material, with a capacity of 285 ml, duly identified), placed on a metal shelf suitable for tubes, with a height of 37.0 cm, having 252 cells, reaching the end of the experiment in the seedling formation phase.

All the tubes were filled with the substrates and previously brought to field capacity with the water of the respective treatment on the day before sowing, which was carried out with 3 seeds per tube, and from then on irrigation was carried out twice a day, in the early morning and late afternoon, so that the tubes remained at field capacity, regardless of the different types of substrates.Then the seedlings were thinned out, leaving only one seedling per container - the one that was the most vigorous (tallest, greenest leaves and no signs of attack by pests and/or pathogens or the most upright) compared to the others, which was perceptible to the naked eye.

The treatments corresponded to 6 combinations of substrate: 100% commercial substrate (SC) - S1, 100% soil (S) - S2, 100% coconut fiber (CF) - S3 and 50% of (S) mixed with 50% of (CF) - S4; 50% of (SC) mixed with 50% of (CF) - S5 and 50% of (SC) mixed with 50% of (S) - S6 and 2 types of water: A1 - supply water and A2 - waste water.

The commercial substrate used in the experiment was composed of simple superphosphate, potassium nitrate, peat, vermiculite and pine bark; the soil - NeossoloRegolitico Distròfico was acquired from the district of Sao José da Mata, PB and the coconut fiber, which is an agro-industrial residue, has 37% humidity, electrical conductivity equal to 0.3 mS.$_{cm-1}$, pH equal to 5.9 from Paraipaba - CE.

Irrigation began shortly after sowing and continued until the end of the experiment, with the help of a 500 ml pissette at a rate of 15 ml per tube, every day. For this purpose, local water supply (A1) was used - from the Companhia de Agua e Esgotos da Paraiba (CAGEPA), located in the municipality of Campina Grande, PB - and wastewater (A2), from the Córrego de Monte Santo, treated by the UASB anaerobic reactor, a wastewater treatment unit that is capable of removing pathogens and suspended solids, considerably reducing the polluting potential of sewage after treatment; so that they could constantly be at or near field capacity (FC).

For this purpose, soil and water analyses were carried out at the Irrigation and Salinity Laboratory belonging to the Department of Agricultural Engineering at the Federal University of Campina Grande, using the methodology presented by EMBRAPA (1997) for soil and APHA (1997) for water, according to Tables 2 and 3, respectively.

Table 2. Physical and chemical characteristics of the soil - Neossolo Regolitico Distròfico.

____

Soil characteristics

Physics	
Textural classification	Clayey loam
Apparent Specific Mass - 33kPa (kg dm)^3	1,45
Porosity (%)	42,35
Field Capacity (g kg)$^{-1}$	83,6
Wilting Point (g kg)$^{-1}$	22,9
Available Water (g kg)$^{-1}$	60,7
Chemistry	
Assortative Complex (cmolc kg)$^{-1}$	
Calcium (Ca)$^{2+}$	1,87
Magnesium (Mg)$^{2+}$	1,05
Sodium (Na)$^+$	0,06
Potassium (K)$^+$	0,23
Saturation Extract (mmol$_c$ L)$^{-1}$	
Cl$^-$	3,75
CO$_3^{2-}$	Absent
HCO$_3^-$	1,70
SO$_4^{2-}$	Present
Ca^{2+}	1,75
Mg^{2+}	2,00
In$^+$	1,12
K$^+$	0,55
PH$_{ps}$	6,15
ECes (dS m)$^{-1}$	0,67

Table 3. Chemical analysis of the water used for irrigation.

Month	pH	Cea (dS.m)$^{-1}$	P- Total	K	N- Total	In	Ca	Mg	Zn	Cu	Fe	Mn	RAS (mmolL 1)$^{0.5}$
					mgL^{-1}								
					Water Supply								
Average	7,2	0,8	a	5,39	a	35,54	20	15,2	a	a	a	a	1,44
					Treated Wastewater								
January	7,6	1,06	3,52	30,36	28,5	171,9	50,1	45,1	0,01	0,008	0,001	0,003	4,51
February	7,7	1,1	3,59	30,42	29,2	171,5	50,9	45,7	0,01	0,006	0,001	0,001	4,22
March	7,9	1,2	3,68	30,47	31,2	178,1	52,3	46,6	0,01	0,006	0,001	0,007	4,26
Average	7,73	1,12	3,60	30,42	29,63	173,83	51,10	45,80	0,01	0,007	0,001	0,004	4,33

a: absence

The experimental design adopted was a randomized block design with a factorial scheme of 6 x 2, with 3 replications, totaling 36 plots, each consisting of three tubes, for a total of 108 experimental units.

Growth evaluations were carried out weekly over a period of 45 days of experimentation, totaling 5 evaluations. The variables analyzed were plant height (AP) (measured from the ground level to the last node of the central stem), number of leaves (NF) and stem diameter (DC) (measured close to the ground), starting 15 days after sowing (DAS).

Phytomass evaluations were carried out at three different times, at 15, 30 and 45 DAS, assessing root length (RC), fresh phytomass of the aerial part (FFPA) and root (FFR), dry phytomass of the aerial part (FSPA) and root (FSR), The stems of each plant were cut off close to the ground using a stylus, and the fresh weight of the aerial part and root were immediately measured using an analytical balance with a precision of 0.01 g. After weighing the fresh phytomass, they were placed in duly identified paper bags and dried in a forced-air oven at 65 °C until they reached a constant weight, determining the dry phytomass of the plants using an analytical balance accurate to 0.0001 g.

The results were evaluated by analysis of variance and the means were compared using the Tukey test at 5% significance, with the aid of the SISVAR statistical software (FERREIRA, 2003).

RESULTS AND DISCUSSION

Table 4 shows the summary of the analysis of variance for the five evaluation periods for the plant height (PH) variable. A statistical difference at the 5% probability level was only perceived for the Type of water factor at 15 (API) and 45 (AP5) days after sowing, however, there were higher AP values for sunflower seedlings irrigated with wastewater in the last three evaluations (Figure 1A), reaching a height of 17.59 cm at the end of the experiment, compared to those irrigated with supply water, which had an average height of 14.19 cm. Souza *et al.* (2010), working with wastewater and organic fertilizer in sunflower cultivation, also found a significant effect ($p < 0.01$) for the water type factor on the variables plant height, stem diameter and number of leaves at 14 and 28 days after transplanting (DAP).The positive effect on plant height with the use of wastewater in agriculture was also observed in other species, in E. *citriodora* plants (FREIER *et al.*, 2007) and in corn plants (COSTA *et al.*, 2009).

For the substrate type factor (S), there was a significant effect ($p < 0.01$) in all evaluation periods.0.01) was found in all the evaluation periods, with the commercial substrate + soil (S6) and soil + coconut fiber (S4) standing out as having the highest plant height averages (Table 4), which, when using materials such as commercial substrate and coconut fiber added to the soil, could favor greater growth due to their physical and chemical characteristics, such as greater total porosity, which provides greater water retention and aeration capacity and a greater amount of essential nutrients for the seedlings (MENDONÇA *et al.*, 2003). This similarity in the superiority of the PA averages for the soil + coconut fiber combination was found by Brito *et al.* (2012), in their work evaluating sunflower cultivars grown in different substrates.

The lowest results were obtained with substrates S2 - soil, S3 - coconut fiber and S5 - commercial substrate + coconut fiber, a fact that can be explained by S2 having a low level of nutrients and/or poor aeration and inadequate porosity, and by substrates S3 and S5 being high-porosity materials, for example, which, when used in isolation, lacked the mainly physical characteristics to provide good PA averages.

Table 4 Summary of the analysis of variance for plant height (PH), in 5 evaluation periods, of sunflower seedlings grown in 6 combinations of agricultural substrates and irrigated with 2 qualities of water.

Mean Squares

Variation Factor	GL	AP_1	AP_2^1	AP_{3i}	AP_4^2	AP_5^2
Type of Water (A)	1	1,063*	0.015ns	0.111ns	1.280ns	2,648*
Type of Substrate (S)	5	3,079**	1,117**	1,745**	2,837**	3,639**
A x S interaction	5	0.103ns	0.040ns	0.067ns	0.367ns	0.434ns
Waste	24	0,240	0,125	0,155	0,369	0,506
CV		19,03	14,43	12,90	17,59	18,11

Type of Substrate	Averages (cm)				
Commercial substrate (S1)	2.88ab	6.68ab	9.01bc	11.83bc	15.30bc
Soil (S2)	1,98c	4,48b	6,28c	8,75c	12,20c
Coconut fiber (S3)	1,90c	4,37b	6,90c	8,27c	10,80c
Soil + coconut fiber (S4)	3,32a	8,18a	12.36ab	17.63ab	23.13ab
Commercial substrate + coconut fiber (S5)	2.02bc	4,48b	5,75c	7,95c	10,07c
Commercial Substance + Soil (S6)	3,48a	9,35a	13,37a	18,43a	23,87a

** and * significant at 1 and 5% probability, respectively; ns - not significant. Averages followed by the same letter vertically do not differ by Tukey's test.[1] Variables transformed into root of x. [2] Variables transformed into root of x+0.5.

As with PA, the number of leaves (NF) variable showed a significant effect for the substrate type factor (p < 0.01), with the exception of the last evaluation (NF5) (Table 5). Once again, the commercial substrate + soil (S6) and soil + coconut fiber (S4) combinations stood out, with the highest NF averages and which did not differ from each other statistically.

The lowest NF averages were provided by combinations S1, S3 and S5, similar to those obtained by AP. So that at 45 DAS, the plants grown in commercial substrate + soil (S6) had an average of 4 leaves more than those grown in commercial substrate (S1).Contrary to this result, Figueiredo *et al.* (2008), in their work with sunflower seedlings, obtained better averages in parameters such as aerial part, fresh weight, dry weight, plant height and number of leaves with the use of commercial substrate (alone), showing that the use of these various substrates independently or in combination can produce significant differences in the growth of sunflower plants.

No significant difference was found for the interaction between the A x S factors and the Water factor for this variable, as shown in Table 5. However, the regression equations presented in Figure 1B show a slight superiority in the number of leaves for the seedlings irrigated with wastewater compared to the water supply at 37 and 45 DAS, NF4 and NF5, respectively. Thus, although there were no differences in the averages, there are numerous advantages to using wastewater for agricultural purposes, such as the possibility of supplying and recycling nutrients, contributing to the preservation of the environment (VAN DER HOEK *et al.*, 2002).

Table 5 Summary of the analysis of variance for the number of leaves (NF), in 5 evaluation periods, of sunflower seedlings grown in 6 combinations of agricultural substrates and irrigated with 2 qualities of water.

Mean Squares

Cause of Variation	GL	NF1	NF2	NF3	NF4[1]	NF5i
Water Type (A)	1	0.18ns	0.44ns	0.11ns	0.12ns	0.59ns
Type of Substrate (S)	5	3,43**	7,61**	10,71**	1,11**	0.69ns
A x S interaction	5	0.16ns	0.01ns	0.04ns	0.23ns	0.22ns
Waste	24	0,24	0,32	0,49	0,21	0,31

CV	17,81	12,54	13,30	18,99	21,29
Type of Substrate	**Averages (unit)**				
Commercial substrate (S1)	2,53b	3,75b	4,41c	4,20b	5,03a
Soil (S2)	2,30b	4,50b	5,83b	6.00ab	7,00a
Coconut fiber (S3)	2,07b	3,75b	3,75c	3,83b	5,50a
Soil + coconut fiber (S4)	3,67a	5,58a	6,58a	8,33a	8,67a
Commercial Substrate + Fiber coconut (S5)	2,26b	3,50b	3,67c	4,00b	5,33a
Commercial Substance + Soil (S6)	3,79a	6,25a	7,41a	8,33a	9,00a

** and * Significant at 1 and 5% probability, respectively; ns - not significant. Averages followed by the same letter vertically do not differ according to Tukey's test. [1] Variables transformed into root of x + 0.5.

With regard to the stem diameter (CD) variable, there was a significant effect at the 1% probability level for the Type of substrate factor in all the evaluation periods (Table 6). Once again, the use of the combination of soil + coconut fiber (S4) and commercial substrate + soil (S6) stood out, with the highest mean stem diameter compared to the other substrates (S1, S2, S3 and S5). The greatest disparity in averages occurred at 30 DAS (DC3), when the seedlings grown in S6 reached an average of 0.28 mm and those grown in S5 only reached 0.13 mm. As these two combinations have the commercial substrate in common, the problem was possibly due to the addition of coconut fiber, as this combination (S5) was the one with the lowest average DC, a fact that was explained in the discussions on the AP variable, which also had a low value.

According to Biscaro *et al.* (2002), high stem diameters in sunflower are considered a desirable characteristic because they make the crop less vulnerable to lodging and because they make it easier to carry out handling and cultivation practices.

Table 6. Summary of the analysis of variance for stem diameter (SD), in 5 evaluation periods, of sunflower seedlings grown in 6 combinations of agricultural substrates and irrigated with 2 qualities of water.

		Mean Squares				
Cause of Variation	**GL**	DC1	DC2	DC3	DC4	DC5
Water Type (A)	1	0.0003ns	0.0001ns	0.0001ns	0.0015ns	0.0034ns
Type of Substrate (S)	5	0,0019**	0,0113**	0,0174**	0,0210**	0,0236**
A x S interaction	5	0.0002ns	0,0005*	0.0003ns	0.0016ns	0.0026ns

Waste	24	0,0002	0,0001	0,0004	0,0016	0,0018
CV		17,57	8,58	11,59	21,06	21,41
Type of Substrate		**Averages (mm)**				
Commercial substrate (S1)		0,0817c	0,1300b	0.1600b	0,1600b	0,1700b
Soil (S2)		0,0800c	0,1350b	0,1700b	0,1667b	0,1883b
Coconut fiber (S3)		0.0900bc	0,1250b	0,1367b	0,1450b	0,1400b
Soil + coconut fiber (S4)		0,1200a	0,2150a	0,2808a	0,2650a	0,2833a
Commercial Substrate + Fiber coconut (S5)		0,0832c	0,1241b	0,1333b	0,1391b	0,1541b
Commercial Substance + Soil (S6)		0.1147ab	0,2100a	0.2808a	0,2667a	0,2775a

** and * significant at 1 and 5% probability, respectively; ns - not significant. Averages followed by the same letter vertically do not differ according to Tukey's test. [1] Variables transformed into root of x + 0.5.

There was no significant effect of the interaction between the factors Type of water and Type of substrate on any of the growth variables studied (AP, NF, DC), with the exception of DC2, which indicates the independence of the factors.

In line with the NF variable, there was a non-significant effect for the Type of water factor on the stem diameter (CD) variable (Table 6), even so, there were higher averages of stem diameter (CD) for seedlings irrigated with wastewater, with an increase of around 2.48%, 7.1% and 10.1% at 30, 37 and 45 DAS over those irrigated with supply water, according to the regressions in Figure 1C, confirming the result found by Andrade *et al.* (2007), in their work on the initial growth of sunflower seedlings, who observed that seedlings irrigated with wastewater obtained a higher number of leaves (NF) and stem diameter (DC) than those irrigated with supply water.Souza *et al.* (2010) also obtained greater increases with the use of wastewater compared to supply water for this same variable.

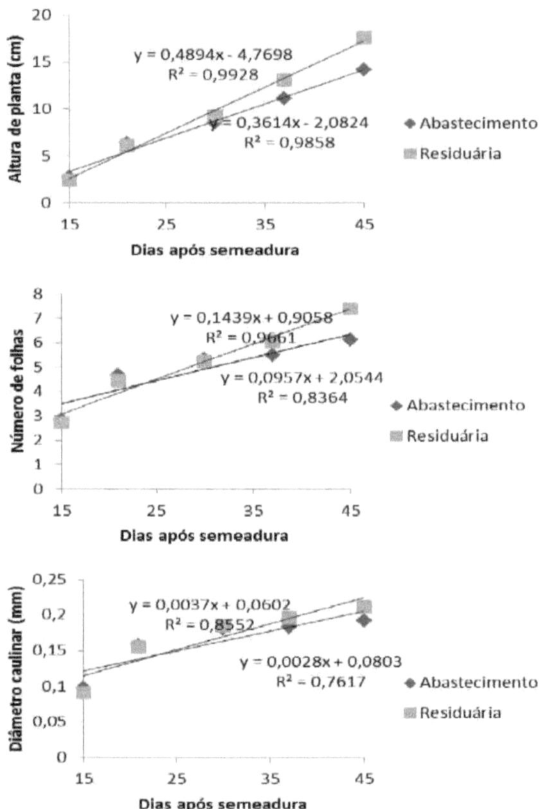

Figura 1. Regressions for plant height (A), number of leaves (B) and stem diameter (C) of sunflower seedlings irrigated with two qualities of water.

According to Souza *et al.* (2010), the best performances with the use of wastewater for the variables DC, AP and NF are probably largely due to the availability of nutrients for the plants, especially the nitrogen (N) present in the wastewater. This nutrient is the one that most limits sunflower production (BLAMEY *et al.*, 1997), causing a reduction of up to 60% in productivity as a result of its deficiency (SMIDERLE *et al.*, 2003). In addition, wastewater contains phosphorus and potassium, which also contribute to the good development of sunflower plants.

Figure 2 shows the regression for root length (RC). The trend lines show higher average RC for the seedlings irrigated with wastewater compared to those irrigated with tap water, a fact that can be attributed to the presence of organic matter, which is important for the soil-plant system (STEVENSON, 1994) and the nutrients contained in domestic sewage. In the first thinning (15 DAS), the seedlings of the two treatments did not differ statistically, as they obtained the same average CR, around 21.0 cm; however, in the 2nd and 3rd thinning (30 and 45 DAS), there were greater root lengths: around 23.0 cm and 27.0 cm, with the use of wastewater, as opposed to 21.0 cm and 23.0 cm for the seedlings irrigated with tap water. Compared to the other variables (AP, NF, DC) studied so far in this work, it is possible to highlight that the use of wastewater improved the growth of the seedlings, although, in most cases, not significantly, which suggests that it should be passively used in agriculture, avoiding its improper disposal in watercourses and favoring the supply of nutrients, reducing fertilization costs.

The same figure shows the mean test for the RC of sunflower seedlings, according to the Type of substrate factor (Figure 2). There was a significant effect only at 15 DAS, with the highest RC averages for the S3 and S5 combinations, with approximately 23.0 cm, unlike the values found for the other growth variables studied. S1 did not differ from S6, nor did S2 differ from S4. These results highlight the importance of studying the most varied combinations of substrates in order to see which ones favor better plant growth and development in general.

In the following evaluations, there were no significant differences, although the seedlings that were thinned out at 30 DAS produced 25.0 cm of CR with the commercial substrate (S1) and 19.0 cm of CR with the soil (S2), the worst average. The seedlings that were thinned on the last evaluation date (Figure 2) produced 28.0 cm when grown in S6 and the worst average was 21.0 cm of root length with S2.

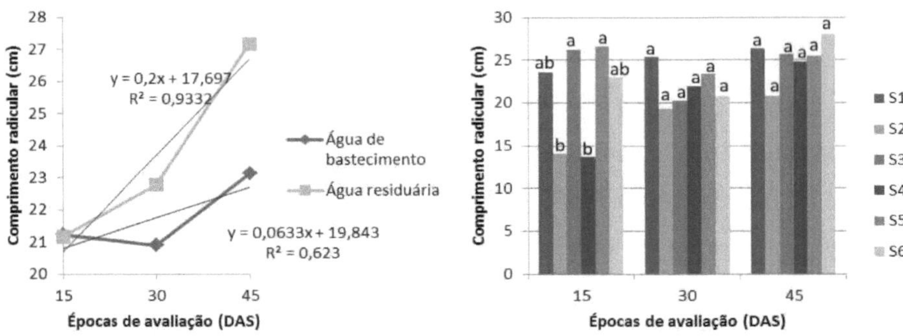

Figura 2. Regression and test of means for the root length of sunflower seedlings grown in six different substrates, irrigated with two qualities of water, in the three evaluation periods. Commercial substrate (S1), Soil (S2), Coconut fiber (S3), Soil + Coconut fiber (S4), Commercial substrate + Coconut fiber (S5) and Commercial substrate + Soil (S6). Averages with the same letter do not differ by the

Tukey.

The fresh phytomass of the aerial part (FFPA), the total fresh phytomass (FFT), the dry phytomass of the aerial part (FSPA) and the dry phytomass of the root (FSR) were not significantly influenced by the type of water, as shown in Table 7. This result agrees with Silva *et al* (2012), who, in their work on sunflower growth irrigated with wastewater, following the behavior of the other vegetative variables, the dry mass of the aerial part did not show statistical significance on the different evaluation dates; This behavior allows us to infer that, although the wastewater used in irrigation did not affect the vegetative variables, it could make it possible to substitute fresh water and save fertilizers in the fertilization, as well as acting as an environmental remedy, avoiding its excess disposal in water bodies. In addition to this advantage, Azevedo and Oliveira (2005) also highlight the importance of using domestic wastewater to supply nutrients and increase plant productivity.

The interaction between the factors was only statistically different for FSPA at 15 DAS and FSR at 30 DAS.

Table 7. Summary of the analysis of variance for the fresh phytomass of the aerial part

(FFPA), fresh phytomass of the root (FFR), dry phytomass of the aerial part (FSPA), dry phytomass of the root (FSR) in 3 evaluation periods, of sunflower seedlings grown in 6 combinations of agricultural substrates and irrigated with 2 qualities of water.

Mean Squares

Variation Factor	GL	FFPAi[1]	FFPA2	FFPA3[3]	FFRi[2]	FFR2[2]	FFR3[3]
Water Type (A)	1	0.0004ns	0.46ns	0.21ns	0,00005 ns	0,0001 ns	0.049ns
Type of Substrate (S)	4	0,2799**	6,04**	0,57**	0,088*	0,2859 **	0.155ns
A x S interaction	4	0.02ns	0.26ns	0.25ns	0.01ns	0,0408 ns	0.230ns
Waste	25	0,13	0,24	0,11	0,02	0,04	0,10
CV (%)		18,79	12,81	20,89	16,26	15,44	21,81

Mean Squares

Cause of Variation	GL	FSPAi[1]	FSPA2[1]	FSPA3[2]	FSRi[2]	FSR2[1]	FSR3[2]
Water Type (A)	1	0,00003ns	0.0081ns	0.0046ns	0,0002ns	0,0009 ns	0,0023 ns
Type of Substrate (S)	5	0,0209**	0,1629**	0,0760**	0,0007*	0,0318 **	0,0101 **
A x S interaction	5	0,0028*	0.0040ns	0.0084ns	0,0003	0,0022 *	0,0027 ns
Waste	24	0,0008	0,0046	0,010	0,0002	0,0013	0,0020
CV (%)		17,69	21,51	11,71	2,09	14,70	5,74

** and * significant at 1 and 5% probability, respectively; ns - not significant. Averages followed by the same letter vertically do not differ by Tukey's test.[1] Variables transformed into root of x;[2] Variables transformed into root of x + 0.5.[3] Variables transformed into root of x + 1.0

On the other hand, the Type of substrate factor had a significant effect on most of the evaluation dates. Figure 3A shows the test of means for FFPA, which had a significant effect on the three evaluation dates at the 1% probability level (Table 7). S4 and S6 had the highest fresh mass averages, a fact that occurred repeatedly in the growth variables studied, with the other genotypes not differing at 15 and 30 DAS.

For the FFR there was a significant effect for the first and second evaluations, as shown in the mean test for the substrate type factor in Figure 3C, with S3 having the highest mass at 15 DAS and S4, followed by S6 at 30 and 45 DAS.

FSPA and FSR showed a significant effect (p<0.001) for the S factor in the three evaluation periods and their average tests can be seen in Figure 3B and 3D, respectively, with the S4 and S6 combinations standing out with the highest dry mass averages at 30 and 45 DAS. On the last evaluation date, there was the greatest disparity for both FSPA and FSR in S6 compared to S5, which had the worst averages for dry phytomass in general, since the seedlings that were thinned at 45 DAS grown in S6 had 0.52 g and 0.21 g compared to 0.05 g and 0.029 g for those grown in S5, in FSPA and FSR, respectively.

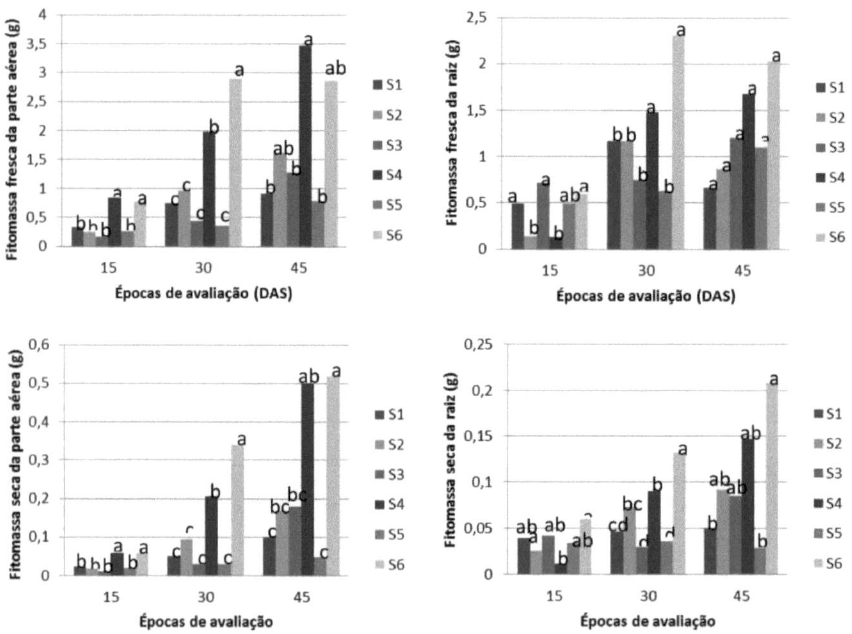

Figure 3 Mean test for fresh (A) and dry (B) phytomass of the aerial part, fresh (C) and dry (D) phytomass of the root, of sunflower seedlings grown in six combinations of agricultural substrates and irrigated with two qualities of water. Commercial substrate (S1), Soil (S2), Coconut fiber (S3), Soil + Coconut fiber (S4), Commercial substrate + Coconut fiber (S5) and Commercial substrate + Soil (S6). Averages with the same letter do not differ according to Tukey's test.

The quantification of dry matter is important information, as sunflower can be used for green manure and silage production (CASTRO and FARIAS, 2005).

CONCLUSIONS

1.   The combinations of commercial substrate + soil and soil + coconut fiber offered higher averages for all the growth variables analyzed, as well as for the main variables related to forage production (FFPA, FFR, FSPA, FSR) in most of the evaluation seasons.

2.   The use of wastewater, although not significantly, provided higher values of PA, NF, DC and CR in most of the evaluation periods, compared to the use of supply water.

3.   Although the wastewater used for irrigation did not affect the vegetative variables, this could make it possible to replace drinking water and save on fertilizers by using reuse water as the main water resource for irrigating sunflower seedlings.

4.   Wastewater promoted considerable increases in sunflower growth and phytomass.

BIBLIOGRAPHICAL REFERENCES

ANDRADE, L.; NOBRE, R.G.; SOARES, F.A.L.; GHEYI, H.R.; FIGUEIREDO, G.R.G.; SILVA, L.A. Germination and initial growth of sunflower plants (Helianthusannuus L.) irrigated with wastewater. Higher Agricultural Education, v.22, n.02, p.48-50, 2007.

APHA - American Public Health Association. Standard methods for the examination of water and wastewater. 20 ed. New York. APHA, AWWA, WPCR, 1997, 1994p.

AZEVEDO, L. P. de; OLIVEIRA, E. L. de. Effects of the application of sewage treatment effluent on soil fertility and cucumber productivity under subsurface irrigation. Engenharia Agricola, v. 25, n. 01, p. 253-263, 2005.

BISCARO, G. A.; MACHADO, J. R.; TOSTA, M. da S.; MENDONÇA, V.; SORATTO, R. P.; CARRIJO, O.A; LIZ, R.S; MAKISHIMA, N. 2002. Green coconut shell fiber as an agricultural substrate. Horticultura Brasileira 20: 533-535.

BLAMEY, F. P. C. et al. Sunflower production and culture. In: SEITEER, A. A. (Ed.).

Sunflower technology and production. Madison: American Society of Agronomy, 1997. v. 1, Chap. 5, p. 595-670.

BRITO, K.S.A.; SILVA, V.F.; SILVA, E.L.; BRITO, B.M.A.; BARACUHY, J.G.V.Growth of sunflowers grown in substrates from solid agro-industrial waste. National meeting of education, science and technology - ENECT. UEPB. Campina Grande, PB. 2012.

CARRIJO, O.A; LIZ, R.S; MAKISHIMA, N. 2002. Green coconut shell fiber as an agricultural substrate. HorticulturaBrasileira 20: 533-535.

CASTRO, C.; FARIAS, J. R.B. Sunflower ecophysiology. In: Leita, R. M. V. B. C.; Brighenti, A. M.; Castro, C. (ed). Sunflower in Brazil. Londrina: EMBRAPA, 2005. p.163-218.

COSTA, F. X. et al. Residual effects of the application of biosolids and irrigation with wastewater on corn growth. Revista Brasileira de EngenhariaAgricola e Ambiental, Campina Grande, v. 13, n. 6, p. 687-693, 2009.

EMBRAPA - Brazilian Agricultural Research Corporation. Manual e métodos de análise de solo. 2 ed. Rio de Janeiro. National Soil Research Center. 1997, 247p.

FERREIRA, D.F. Programa Sisvar - statistical analysis program. Lavras: UFLA, 2003.

FIGUEIREDO, G.R.G.; ANDRADE, L.O.;BATISTA, D.S.; FARIAS, G.A.; NOBRE, R.G.; RÊGO, E.R. Production of ornamental sunflower seedlings (Helianthusannuus L. cv. Dobrado Sungold) in different substrates. Revista Educaçao Agricola Superior. Associaçao Brasileira de Educaçao Agricola Superior- ABEAS - v.23, n.1, p.105-107, 2008.

FREIER, M.; MALAVASI, U.C.; MALAVASI, M.M. Effects of the application of biosolids on the initial growth of EucalyptuscitriodoraHook. 2007.

MENDONÇA, V.; ARAÙJO NETO, S.E; RAMOS, J.D.; PIO, R.; GONTIJO, T.C.A. Different substrates and containers in the formation of "Sunrise Solo" papaya seedlings. Revista Brasileira de Fruticultura, v.25, n.1, p. 127-130, 2003.

NASCIMENTO, N. O.; HELLER, L. Science, technology and innovation at the interface between the areas of water resources and sanitation. Engenharia Sanitària eAmbiental, Rio de Janeiro, v. 10, n. 1, p. 36-48, 2005.

OLIVEIRA, A.A.P.; BRAINER, M.S.C.P. Floriculture: characterization and market. Banco do Nordeste, Fortaleza, Série Documentos do ETENE, n. 16, p.180, 2007.

ROSA, M.F.;BEZERRA, F.C.; CORREIA, D.; SANTOS, F.J.S.; ABREU, F.A.P.; FURTADO, A.A.L.; BRiGIDO, A.K.L.; NORÔES, E.R.V Utilization of coconut husk as an agricultural substrate. Embrapa Agroindustria Tropical. Documentos, 52. Fortaleza, 24 p., 2002.

SALUNKHE, D. K.; DESAI, B. B. Sunflower.In: SALUNKHE, D. K.; DESAI, B. B. Postharvest biotechnology of oilseeds. Boca Raton: CRC Press, 1986. p.57-92.

SILVA, L.T.; OLIVEIRA, M.L.A.; SACRAMENTO, D.S.; MENDONÇA, J.O.G.; OLIVEIRA, X.S.; GHEYI, H.R. Sunflower growth irrigated with treated wastewater. 2012. Available at:< http://www.inovagri.org.br/meeting/wp-content/uploads/2012/06/Protocol418.pdf> Accessed on February 28, 2013.

SMIDERLE, O. J. et al. Nitrogen fertilization, spacing and sowing times for sunflower in the Cerrados of Roraima. In: EMBRAPA. EMBRAPA Soja research results, 2003. p. 33-39. (Documents, 218).

SOUSA, J.T.; CEBALLOS, B.S.O.; HENRIQUE, I.N.; DANTAS, J.P.; LIMA, S.M.S. Reuse of wastewater in the production of pepper (Capsicumannuum L.). Revista Brasileira de Engenharia Agricola e Ambiental, v.10, p.89-96, 2006.

SOUZA F.X. 2001. Materials for formulating substrates in the production of seedlings and the cultivation of potted plants. Fortaleza: Embrapa- CNPAT. 21p. (Documents 43).

SOUZA, R.M.; NOBRE, R.G.; GHEYI, H.R; DIAS, N.S.; SOARES, F.A.L; Use of wastewater and organic fertilizer in sunflower cultivation. RevistaCaatinga, v.23, n.2, p. 125-133, 2010.

STEVENSON, J.F. Humus chemistry genesis: Composition, reactions. 2.ed. New

York: John Wiley & Sons, 1994. 496p.

UNGARO, M. R. G. Culturadogirassol. Campinas: Agronomic Institute, 2000. 36p. (Technical Bulletin, 188).

VAN DER HOEK, W.; HASSAN, U.M.; ENSINK, J.H.J.; FEENSTRA, S.; RASCHID-SALLY, L.; MUNIR, S.; ASLAM, R.; ALIM, N.; HUSSAIN, R.; MATSUNO, Y. Urban wastewater: a valuable resource for agriculture. A case study from Horoonabad, Pakistan. Colombo, Sri Lanka: International Water Management Institute, 2002. 29 p. (Research Report, 63).

CHAPTER 2

PRODUCTION OF OLEAGINOUS SEEDLINGS IN VARIOUS SUBSTRATES FROM AGRO-INDUSTRIES

ABSTRACT: Agro-industries generate waste that can be reused to minimize environmental impacts. In this context, the work was carried out in a protected environment at the Federal University of Campina Grande, with the aim of evaluating the quality of oleaginous sunflower seedlings produced in various substrates made from agro-industrial waste. The materials used to compose the mixtures consisted of sand and industrial waste such as coconut fiber and sugar cane bagasse. Four sunflower cultivars were used in the experiment: AG 262, Olisun 3, Helio 253 and Embrapa 122 - 2000. The ratio of plant height to crown diameter, relative and absolute growth rate of plant diameter and height, total dry mass, the ratio of aerial part to root system and the Dickson quality index were evaluated. Embrapa 122-2000 had the highest average absolute growth rate of 9.77 cm/day. In relation to the weight of the dry mass of the aerial part and the weight of the dry mass of the root, the substrates had values ranging from 1.43 to 3.33 and the cultivars ranged from 1.40 to 2.54. Coconut fiber and the cultivars Embrapa 122-V200, AG 262 and Olisun 3 performed well.

Keywords: Waste, sunflower, coconut fiber, sugar cane bagasse.

INTRODUCTION

With the increase in food production, agricultural systems have become more intensive with the modernization of agriculture. In this way, a modern industrial segment emerged, responsible for processing primary food production, the agro-industry. According to Spadotto & Ribeiro (2006), the agro-industry is one of the main segments of the Brazilian economy, with importance in both domestic supply and exports.

The problems currently causing concern in the agro-industrial sector are the generation of waste and its final destination. In these activities, the waste generated has the potential to impact on the environment if not treated properly. The environmental impacts associated with these wastes stem from their high generation in quantitative terms and slow degradability in some cases, and in others, from the generation of by-products that can be toxic, cumulative or difficult to degrade (IPEA, 2012).

Once generated, waste needs to be properly disposed of because, as well as creating potential environmental problems, waste represents a loss of raw materials and energy, requiring significant investment in treatments to control pollution. The food industry produces several residues with a high reuse value (PELIZER *et al.*, 2007).

According to Matos (2005), rice husks, for example, account for an average of 20 to 25% of the weight of the grain, while around 39% of the weight of coffee fruit is made up of the husk. Sugar mills and distilleries produce sugarcane bagasse (the residue from sugarcane milling) and filter cake (the residue obtained after filtering the sugarcane juice) as solid waste, so that processing 1000 tons of sugarcane yields an average of 280 tons of bagasse (JOÃO, 2009) and 35 tons of filter cake.

The use of solid waste from agricultural and agro-industrial processes prevents the accumulation of waste, improves pollution control, improves public health conditions, reduces dependence on imported chemical fertilizers and enables sustainable development in agriculture. The use of this waste as fertilizer or substrate allows for the recovery of elements present in the waste, such as nitrogen (N), phosphorus (P), potassium (K) and trace elements. In addition, the addition of organic matter to the soil helps to improve its physical structure, water absorption capacity and the supply of nutrients to plants, making it possible to increase production and improve food quality (MALHEIROS and PAULA JÙNIOR, 1997).

In order for a material to be used as a substrate for seedlings, Severino *et al.* (2006) state that as well as having the appropriate chemical and physical characteristics, it must be available close to the production site in sufficient quantity, as well as being low cost; generally, residues from agro-industries or agricultural processes meet these requirements, such as peanut or castor bean shells, sisal mucilage, ash, sugar cane bagasse, oil extraction cake, etc.

Growing plants in alternative substrates has been increasingly used in our country. The sunflower has special characteristics in terms of its economic potential, as it is very dynamic in terms of the different ways in which it can be used, as well as being a plant well known for its ornamental potential. Sunflower can be used in a variety of ways, including medicinal use, green manure, crop rotation, beekeeping, animal feed, oil production, human food and biodiesel production (SILVA *et al*, 2011).

According to Uchôa *et al.* (2011), the sunflower crop has one of the highest growth rates in the world due to its wide adaptability, high drought tolerance and high grain and oil yields. This is also justified by the fact that the sunflower plant, the grains, the remains of the crop and the by-products generated from extracting the oil are used in animal feed.

The aim of this study was to evaluate the production of oleaginous sunflower seedlings grown in different substrates made from agro-industrial waste.

MATERIAL AND METHODS

The experiment was carried out in a protected environment, belonging to the Academic Unit of Agricultural Engineering (UAEAg), linked to the Center for Technology and Natural Resources (CTRN) of the Federal University of Campina Grande - UFCG, located in the municipality of Campina Grande, State of Paraiba-PB, at geographic coordinates 7°15'18" south latitude and 35°52'28" west longitude, at an altitude of 550 m.

The treatments tested resulted from the following volumetric combinations of materials:

S1)Coconut fiber (100%)

S2)Coconut fiber (50%) + sand (50%)

S3) Sugar cane bagasse (100%)

S4)Sand (100%)

Four (4) varieties of sunflowers were used in the experiment:

V   1) AG 262

V   2) Olisun 3

V   3) Helio 253

V   4) Embrapa 122 - V2000

The sunflower varieties were sown and grown in tubes (black plastic with a capacity of 285 ml, duly identified), placed on a metal rack suitable for tubes, 37.0 cm high, with 252 cells, reaching the end of the experiment at the seedling stage.

All the tubes were filled with the substrates and 3 seeds were sown per tube. Irrigation was carried out in the late afternoon with a variable volume in order to always leave the tubes at field capacity, regardless of the different types of substrates and sunflower varieties.

Thinnings were made 15 days after sowing (DAS), leaving only the plant that showed the most vigor (greatest height, greenest leaves and no signs of attack by pests and/or pathogens or most upright) in relation to the others, perceptible to the naked eye.

The experimental design adopted was a randomized block design, in a 4 x 4 factorial scheme, with 3 replications, totaling 48 plots, each consisting of 2 tubes, for a total of 96 experimental units.

The first growth assessment took place at 15 DAS, 21 DAS and 30 DAS, and assessed the plant height/collector diameter ratio (RAD), obtained from the ratio between the height of the aerial part and the diameter of the collar; relative growth rate (TCR) and absolute growth rate (TCA) of the diameter (D) and plant height (A). The formulas used to determine absolute and relative growth were (BENINCASA, 2003), ACR = (V2 - V1)/(T2 - T1) and RCT = (ln(V2) - ln(V1))/(T2 - T1). Where, according to the formula indicated, V stands for the variable and T is the time of each period, respectively.

For the quality of the seedlings in terms of fresh and dry phytomass of the aerial part and root system, the following characteristics were determined at the end of the experiment at 30 DAS: the total dry mass (MST), obtained from the sum of the dry mass of the aerial part and the dry mass of the root system; the aerial part/root system ratio (RPAR), obtained from the ratio between the dry mass of the aerial part and the dry mass of the root system; and the Dickson quality index (IQD), according to Dickson et al, (1960), where DQI = MST/(RAD+RPAR). The RAD used in the DQI was from the last growth assessment.

The height of the seedlings was determined from the neck of the plants to the meristematic apex and the cotyledonary leaves were excluded. The stems of each plant were cut off close to the ground and the fresh weight of the aerial part and the root were immediately weighed using an analytical balance accurate to 0.01g. After weighing the fresh phytomass, they were placed in duly labeled paper bags and dried in a forced-air oven at 65 °C until a constant weight was obtained, and the dry phytomass of the plants was determined using an analytical balance accurate to 0.0001g.

The results were evaluated by analysis of variance and the means were compared using the Tukey test at 5% significance, with the aid of the SISVAR statistical software (FERREIRA, 2003).

RESULTS AND DISCUSSION

Table 8 shows that the type of substrate and sunflower variety in RAD 1 and RAD2 showed a statistical difference (p<0.01), but when these variables interacted, it was only significant at 5% in RAD3.

Table 8. Summary of the analyses of variance for the ratio of plant height to crown diameter (RAD) in the three evaluations and the relative (TCR) and absolute (TCA) growth rate of plant height (A) and crown diameter (D) in the first and last evaluations for the sunflower cultivars in different substrates.

Mean Squares[1]								
Source of Variation	GL	RADI	RAD2	RAD3	TCAA	TCRA	TCAD	TCRD
substrate (s)	3	6,74**	4,24**	6.93 ns	47,17**	1,27 **	0,36 **	0,16**
Cultivars (C)	3	11,67**	16,57**	3.75 ns	74,92**	1,07 **	0,25 **	0.02 ns
substrate * Cultivars (s* C)	9	0.24 ns	0.95 ns	6,19 *	1.76 ns	0.09 ns	0.02 ns	0.01 ns
Waste	32	0,34	0,66	2,42	1,28	0,12	0,02	0,01
CV (%)		15,78	15,81	24,05	16,29	19,87	6,54	17,13
Type of substrate				Average				

Coconut fiber (s1)	4,26 c	5,29 b	7,01 a	8,51 c	2,05 b	2,10 b	0,76 b
Coconut fiber + Sand (s2)	4,29 c	5,48 b	6,93 a	8,68 c	2,07 b	2,14 b	0,78 b
Sugar cane bagasse (S3)	3,57b	5,53 b	6,58 a	5,93 b	1,48 a	1,77 a	0,56 a
Sand (s4)	2,70 a	4,26 a	5,37 a	4,64 a	1,50 a	1,89 a	0.66 ab
Variety				Average			
AG 262	4,32 c	6,25 c	7,19 a	8,06 c	2,01 b	1,97 a	0,70 a
Olisun 3	3,33 b	4,98 b	6,01 a	5,86 b	1.72 a b	1,91 a	0,67 a
Helio 253	2,50 a	3,57 a	5,98 a	4,06 a	1,37 a	1,85 a	0,64 a
Embrapa 122-V2000	4,68 c	5.77 bc	6,64 a	9,77 d	1,99 b	2,18 b	0,74 a

[1] significant at 1% (**), at 5% (*) and (ns) not significant by Tukey's test; CV - coefficient of variation; ASCT and ARCT in cm/day; ADCT and ARCT in mm/day.

The absolute and relative growth rate of plant height was significant at 1% and the absolute growth rate of diameter, but the relative growth rate was only significant for the substrates (p<0.01). The highest coefficient of variation for RAD was 24.5%, with no statistically significant effect for the substrate and variety variable (Table 8).

For the ratio of plant height to crown diameter between the different cultivars (Table 8), it can be seen that Embrapa 122-V2000 had the highest average in the first evaluation, and in the subsequent ones (RAD2 and RAD3) the AG 262 cultivar had averages of 6.25 and 7.19, higher than the other sunflower species studied. Analyzing the RAD in relation to the substrate, coconut fiber had the highest averages in the three evaluations with 4.26, 5.29 and 7.01.

Seedlings with small neck diameters and high heights are considered to be stunted and are of lower quality than those with smaller and larger neck diameters. This ratio represents a balance in the growth of these two important parameters in a single index and is known as the "vigor" of the seedlings (CARNEIRO, 1995). The lower its value, the greater the seedlings' ability to survive and establish themselves in the definitive planting area (CARNEIRO, 1983).

As for the ratio of plant height to crown diameter (RAD), all the sunflower species studied had an index of less than 10, the standard recommended by Birchler *et al.* (1998). For *Pinus taeda* seedlings, Carneiro (1995) comments that for a good result for this ratio, the average values should be between 5.4 and 8.1. The sunflower cultivars in different agro-industrial substrates had an excellent RAD, with coconut fiber standing out for all cultivars at 30 DAS.

In terms of absolute and relative growth of the aerial part and diameter in relation to the substrate, coconut fiber combined with sand had the highest averages (Table 8) followed by coconut fiber with very similar values. When analyzing the cultivars in terms of the absolute growth rate of the aerial part, Embrapa 122-2000 had the highest average at 9.77 cm/day, but cultivar AG 262 had a higher average (2.01 cm/day) in terms of the relative growth rate of plant height, with little difference from Embrapa 122-V2000 at 1.99 cm/day. In terms of shoot diameter, the TCAD and TCRD of Embrapa 122-V2000 outperformed the others, followed by AG 262, Olisun 3 and Helio 253.

The value resulting from dividing the height of the plant by its respective collar diameter expresses the balance of growth, relating these two important morphological parameters in just one index (CARNEIRO, 1995), also known as the robustness quotient, which is considered one of the most accurate, as it provides information on how thin the seedling is (JOHNSON & CLINE, 1991).

Junior Maia *et al.*, (2013) carried out studies on the growth rates of sunflower cultivars under different water regimes using BRS Gira 26, Agrobel 962 and Embrapa 122/V2000, found that the cultivar Embrapa122/V2000 had values of 1.31 cm / day and 0.031 mm / day, and for the relative growth rates, values of 0.0974 cm / day and 0.0143 mm / day were found. Compared to the author's data, the results obtained from the Embrapa 122-V2000 sunflower variety in different substrates had better growth development with higher values.

For substrates (Table 9), there was a significant difference (p<0.01) for total dry mass and the Dickson quality index. IQD and MST were significant at 5% for all sunflower cultivars. The coefficient of variation for the ratio of the weight of the dry mass of the aerial part to the weight of the dry mass of the root system was the highest at 96.01 %, with no statistically significant effect for the interaction between cultivars and substrates.

According to Brissette (1984), for the ratio of aerial dry mass weight to root dry mass weight (RPAR), the value that best represents this ratio is equal to 2.0. This ratio is considered to be an efficient and reliable index for expressing the quality standard of these seedlings (PARVIAINEN, 1981). In comparison with the authors, all the results obtained (Table 9) for the substrates were 1.43 to 3.33 and for the cultivars ranged from 1.40 to 2.54, verifying that the Helio 253 cultivar is at an efficient and safe standard.

The sugar cane bagasse substrate had a high RPAR value of 3.33, showing that this substrate is not suitable for sunflower seedlings, so the composition of sugar cane bagasse with other materials needs to be studied in order to improve its performance.

The weight of the dry mass is a good indication of the resilience of seedlings in field conditions, even if it is a destructive method (GOMES, 2001).

Table 9. Summary of the analysis of variance for the total dry mass (MST), the ratio of the weight of the dry mass of the aerial part to the weight of the dry mass of the root system (RPAR) and the Dickson quality index (IQD) for the sunflower cultivars in different substrates.

Mean Squares[1]				
Source of Variation	**GL**	**MST**	**RPAR**	**IQD**
Substrate (S)	3	0,71**	9.33ns	0,30**
Cultivars (C)	3	0,04*	2.70 ns	0,04 *
Substrate * Cultivars (S* C)	9	0.01 ns	3.50 ns	0.02 ns
Waste	32	0,01	3,75	0,01
Type of Substrate				**Average**
Coconut fiber (S1)		0,68 d	1,74 a	0,43 c
Coconut fiber + Sand (S2)		0,57 c	1,57 a	0.37 bc
Sugar cane bagasse (S3)		0,15 a	3,33 a	0,07 a
Sand (S4)		0,30 b	1,43 a	0,25 b
Variety				**Average**
AG 262		0,47 b	2,54 a	0,30 a
Olisun 3		0.43 ab	2,14 a	0,31 a
Helio 253		0,34 a	2,0 a	0,19 a

| Embrapa 122-2000 | 0,45 b | 1,40 a | 0,32 a |

[1]GL - degree of freedom; significant at 1% (**), at 5% (*) and (ns) not significant by the F test; CV - coefficient of variation;

It is important to emphasize this ratio of aerial part dry mass weight to root system dry mass weight (RPAR) when the seedlings go into the field, as the aerial part of the seedlings should not be much larger than the root part, due to problems that may occur in relation to the absorption of water into the aerial part.

The Dickson Quality Index is mentioned as a promising integrated morphological measure (JOHNSON & CLINE, 1991) and is considered a good indicator of seedling quality, as its calculation takes into account the robustness and balance of the biomass distribution in the seedling, weighing up the results of several important parameters used to assess quality (FONSECA *et al.*, 2002).

In this study, the species with the highest Dickson Quality Index, i.e. the best seedling quality, were Embrapa 122- V200, AG 262, Olisun3 and Helio 253, respectively.

CONCLUSIONS

1. Embrapa 122-2000 and AG 262 had greater absolute and relative growth, both in plant height and collar diameter.

2. The species with the best seedling quality were Embrapa 122- V200, AG 262 and Olisun 3, respectively.

3. The sunflower cultivars grown in different agro-industrial substrates had an excellent ratio of plant height to crown diameter (RAD). The Helio 253 cultivar is at an efficient and safe standard when it comes to the weight of the dry mass of the aerial part and the weight of the dry mass of the root.

4. The best substrate to use for seedling production is pure coconut fiber, as well as 50% mixed, with the highest absolute and relative growth averages.

5. The use of agro-industrial waste is a sustainable and viable alternative for producers of seedlings of various species.

BIBLIOGRAPHICAL REFERENCES

BENINCASA, µ. M. P. Plant growth analysis (basics). Jaboticabal, SP, FUNEP, 2003, 41p

BIRCHLER, T.; ROSE, R. W.; ROYO, A.; PARDOS, M. La planta ideal: revision Del concepto, parâmetros definitorios e implementaction practica. Investigacion Agraria, Sistemas y Recursos Forestales, Madrid, v. 7, n. 1/2, p. 109121, 1998.

BRISSETTE, J. C. Summary of discussions about seedling quality. In: SOUTHERN NURSERY CONFERENCES, 1984, Alexandria. Proceedings...New Orleans: USDA Forest Service, Southern Forest Experiment Station, 1984. p.127128.

CARNEIRO, J. G. A. Influence of environmental factors and production techniques on the development of forest seedlings and the importance of the parameters that define their quality. In: FORESTS PLANTED IN THE NEOTROPICS AS A SOURCE OF ENERGY, 1983. Proceedings... Viçosa: Federal University of Viçosa, 1983. p. 10-24.

CARNEIRO, J. G. A. Production and quality control of forest seedlings. Curitiba: UFPR/FUPEF, 1995. 451 p.

DICKSON, A.; LEAF, A.L.; HOSNER, J. F. Quality appraisal of white spruce and white pine seedling stock in nurseries. Forestry Chronicle, v. 36, p. 10-13, 1960.

FERREIRA, D.F. Programa Sisvar - statistical analysis program. Lavras: UFLA, 2003.

FONSECA, E. P.; VALÉRI, S. V.; MIGLIORANZA, E.; FONSECA, N. A. N.; COUTO, L. (2002),Quality standard of Trema micrantha (L.) Blume seedlings, produced under different shading periods. *Revista Arvore*, **26**,515-523.

INSTITUTO DE PESQUISA ECONÔMICA APLICADA - IPEA 2012. Diagnosis of Organic Waste from the Agroforestry Sector and Associated Agroindustries. Brasilia, 2012.

JOÂO, I.S. 2009. Environmental Impact and Waste Management in the Sugarcane Industry: the Case of Using Sugarcane Bagasse to Generate Electricity. Xi Encontro Nacional e i Encontro Internacional sobre Gestão Empresarial e Meio Ambiente.

JoHNsoN, J. D.; CLiNE, p. M. seedling quality of southern pines. in: DuREYA, M. L., DouGHERTY, p. M. (Eds.). Forest regeneration manual. Dordrecht: Kluwer Academic publishers, 1991. p. 143-162.

JuNioR MAiA, s. o.; ANDRADE, J. D.; ARAuJo, D. L.; sousA, J. s.; MEDEiRos, i. F. s. Growth rates of sunflower cultivars under different water regimes. Revista Verde de Agroecologia e Desenvolvimento Sustentâvel Artigo Cientifico, v. 8, n. 3, p. 150 - 155, jul - sep , 2013.

MALHEIROS , S. M. P. &PAULA JÛNIOR, D. R. Utilization of the composting process with agro-industrial waste. *In* : BRAZILIAN CONGRESS OF SOIL SCIENCE, 26, 1997.

MATOS, A. T. 2005. Treatment of agro-industrial waste. Course on the treatment of agro-industrial waste. Federal University of Viçosa. Minas Gerais.

OLIVEIRA, A. B.; HERNANDEZ, F. F. F.; JuNIOR ASSIS, R. N. Green coconut dust, an alternative substrate for eggplant seedling production. Rev. Ciên. Agron., Fortaleza, v. 39, n. 01, p. 39-44, Jan.- Mar., 2008.

PARVIAINEN, J. V. Quality and quality evaluation of forest seedlings. In: SEMINARIO DE SEMENTES E VIVEIROS FLORESTAIS, 1, 1981, Curitiba. Anais...Curitiba: FuPEF, 1981. p.59-90.

PELIZER, L. H.; PONTIERI, M. H.; MORAES, I. de O. Utilization of agro-industrial residues in biotechnological processes as a perspective for reducing environmental impact. Journal of Technology Management & Innovation, v 2, p. 118127, 2007.

SEVERINO, L. S.; LIMA, R. L. S.; BELTRÂO, N. E. M. Composito Quimica de Eleze Materiais Orgânicos Utilizados em Substratos para Produçâo de Seedlings. Ministry of Agriculture, Livestock and Supply. ISSN 0102-0099. August/2006, Campina Grande, PB.

SILVA, A. R. A.; BEZERRA, F. M. L.; SOUSA, C. C. M.; PEREIRA FILHO, J. V.;FREITAS, C. A. S. Performance of sunflower cultivars under different irrigation rates in the Curu Valley, CE. Revista Ciência Agronômica, v. 42, n. 1, p. 57-64, 2011.

SPADOTTO , C.; RIBEIRO , W. Waste management in agriculture and agro-industry. Sâo Paulo: FEFAP, 2006.

UCHÔA, S. C. P.; IVANOFF, M. E. A.;ALVES, J. M. A. A.; MARTINS, T. S.; MARTINS, S. A. Potassium Fertilization in Coverage on the Production Components of Sunflower Cultivars. Revista. Ciência Agronômica, v. 42, n. 1, p. 815, jan- mar, 2011.

CHAPTER 3

GERMINATION OF SUNFLOWER GENOTYPES SUBMITTED TO DIFFERENT SUBSTRATES

ABSTRACT: Reusing agro-industrial waste as a substrate is an alternative that benefits the environment and the producer. In this context, research was carried out at the Federal University of Campina Grande/PB, with the aim of evaluating the effect of different substrates from agro-industrial waste on the germination of sunflower genotypes. The germination percentage (PG), germination speed index (IVG), average speed (Vm) and average time (Tm) of germination of each sunflower variety studied in the different substrates used were evaluated. The design was a randomized block and the analyses were carried out using the computer program System for Analysis of Variance - SISVAR. In relation to the substrate, the germination percentage ranged from 48.17 % to 65.47 %. Helio 253 had the lowest GVI (1.41), while Olisum 3 had the highest (4.80), followed by genotypes AG 262 (4.43) and Embrapa 122- v2000 (3.26). The average germination time (Tm) ranged from 8.0 to 8.92. The substrates considered most suitable for germination of all the genotypes were sand, coconut fiber and coconut fiber + sand.

Keywords : coconut fiber, agro-industrial residue, Embrapa 122-V2000

INTRODUCTION

In Brazil, sunflower cultivation has favorable characteristics due to its adaptability to various regions, resistance to drought, cold and heat and low production costs, making it an alternative for planting in the off-season (BACAXIXI *et al., 2011),* Its productivity is the result of the sum of various plant characteristics and agronomic techniques that interact with each other and with the environment to enable the variety's genetic potential to be expressed (DAROS *et al.*, 1995).

Sunflower cultivation in alternative substrates has been increasingly used. Sunflower can be used in a variety of ways, including medicinal, ornamental, green manure, crop rotation, beekeeping, animal feed, oil production, human food and biodiesel production

(SILVA *et al.*, 2011). Sunflower is a new option for biofuel production (BALBINOT Jr., 2009).

For a material to be used as a substrate for seedlings, Severino *et al.*(2006) state that in addition to having appropriate chemical and physical characteristics, it needs to be available close to the production site in sufficient quantity, as well as being low cost; generally, residues from agro-industries or agricultural processes meet these requirements, such as peanut or castor bean shells, sisal mucilage, ash, sugar cane bagasse, oil extraction cake, etc.

The use of substrates from processes used in agriculture and agro-industry prevents the accumulation of waste, improves pollution control, improves public health conditions, reduces dependence on imported chemical fertilizers and enables sustainable development in agriculture. The use of this waste as fertilizer or substrate allows for the recovery of elements present in the waste, such as nitrogen (N), phosphorus (P), potassium (K) and trace elements. In addition, the addition of organic matter to the soil helps to improve its physical structure, water absorption capacity and the supply of nutrients to plants, making it possible to increase production and improve food quality (MALHEIROS and PAULA JÛNIOR, 1997).

According to the rules for Seed Analysis (BRASIL, 1992), in addition to light, temperature and oxygen, the substrate is of fundamental importance in the results of the germination test. The best substrates should have, among other important characteristics, easy availability for purchase and transportation, absence of pathogens, richness in essential nutrients, adequate pH, good texture and structure (SILVA *et al.*, 2001).

The aim of this study was to evaluate the effect of different substrates made from agro-industrial waste on the germination of sunflower genotypes.

MATERIAL AND METHODS

The experiment was carried out in a protected environment belonging to the Academic Unit of Agricultural Engineering (UAEAg) of the Federal University of Campina

Grande - UFCG, located in the municipality of Campina Grande, state of Paraiba - PB, at the geographical coordinates 7°15'18" south latitude and 35°52'28" west longitude, at an altitude of 550 m.

The treatments tested resulted from the following volumetric combinations of materials: T1) Coconut fiber (100%); T2) Coconut fiber (50%) + sand (50%); T3) Sugarcane bagasse (100%) and T4) Sand (100%).

Four sunflower genotypes were used: G1) AG 262; G2) Olisun 3; G3) Helio 253 and G4) Embrapa 122 - V2000.

The sunflower varieties were sown and cultivated in tubers, duly identified, placed on a metal rack suitable for tubers, with a height of 37.0 cm, with 252 cells, reaching the end of the experiment at the seedling stage.

All the tubes were filled with the substrates and 3 seeds were sown per tube. Irrigation was carried out in the late afternoon with a variable volume in order to always leave the tubes at field capacity, regardless of the different types of substrates and sunflower varieties.

The experimental design adopted was a randomized block design, in a 4 x 4 factorial scheme, with 3 replications, totaling 48 plots, each consisting of 2 tubes, for a total of 96 experimental units.

The germination percentage (PG), germination speed index (IVG), average speed (Vm) and average time (Tm) of germination of each sunflower variety studied in the different substrates used were evaluated.

The germination percentage (PG) was calculated using equation 1, according to Labouriau and Valadares (1976):

$$PG = \left(\frac{N}{A}\right) 100 \qquad \text{Equação (1)}$$

In which:

N - total number of germinated seeds.

A - total number of seeds placed to germinate.

The emergence rate (IVG) was determined by recording the number of germinated seeds daily (Equation 2) up to the 14th day and calculated using the formula proposed by Maguire (1962). Seedlings with completely free cotyledons were considered to have emerged.

$$IVG = \frac{G1}{N1} + \frac{G2}{N2} + \cdots + \frac{Gn}{Nn}$$ Equação (2)

In which:

G_1, G_2 and Gn - number of normal seedlings counted in the first, second and last counts.

N_1, N_2 and Nn - number of days after test implantation.

Average Germination Time (Tm) in days (equation 3) according to Labouriau & Valadares (1976).

$$Tm = (\textstyle\sum ni.\, ti / \sum ni$$ Equação (3)

Where:

ni - number of seeds germinated in a period of time;

and ti - germination time interval.

Using equation 4, the Average Germination Speed (Vm) was calculated (LABOURIAU E VALADARES,1976).

$$Vm = 1/T$$ Equação (4)

Where:

T- Average Germination Time (seeds/day).

The analyses were carried out using the computer program System for Analysis of Variance - SISVAR (FERREIRA, 2003).

RESULTS AND DISCUSSION

Table 9 shows the estimated values and significance of the mean squares, as well as the means and percentage coefficients of experimental variation for the sunflower genotypes in different substrate compositions.

In Table 10, the germination percentage was statistically significant (p<0.01) for the substrate and the genotype, and when the two were compared, the difference was significant at 5%. With regard to the substrate, the germination percentage ranged from 48.17 % to 65.47 %, with the following substrate combinations standing out: sand (65.47 %), sand + coconut fiber (63.46 %) and coconut fiber (61.63 %). With 74.81 %, the Embrapa 122-v2000 genotype had the highest germination percentage, followed by Helio 253 (73.26 %), Olisum 3 (61.44 %) and AG 262 (29.21), with a statistical difference between them. These results agree with those verified by Linhares *et al.* (2005), who, when testing various substrates for sunflower vigor and emergence, verified the best germination rates in sand.

Table 10. Summary of the analysis of variance for germination percentage (PG), germination speed index (IVG), average time (Tm) and average speed (Vm) of germination.

Mean Squares[1]					
Source of Variation	GL	PG	IVG	Tm	Vm
Substrate (T)	3	1719,3223**	16,32**	4.36 ns	0,0013*
Genotypes (G)	3	12552,733**	65,5**	2.54 ns	0,0054**
Substrate * Genotypes (T* G)	9	651,050*	4,24**	1.10 ns	0.00024 ns
CV (%)		25,05	29,34	20,40	15,82
Type of Substrate			Average		
Coconut fiber (T1)		61,63 a	3,63 a	8,05 a	0,116 ab
Coconut fiber + Sand (T2)		63,46 a	3,74 a	8,14 a	0.1152 ab
Sugar cane bagasse (T3)		48,17 b	2,38 b	8,92 a	0,105 b
Sand (T4)		65,47 a	4,14 a	8,27 a	0,121 a
Genotypes			Average		

AG 262	29,21 c	4,43 a	8,2 a	0,122 a
Olisun 3	61,44 b	4,80 a	8,64 a	0,1161 a
Helio 253	73,26 a	1,41 c	8,54 a	0,0944 b
Embrapa 122-V2000	74,81 a	3,26 b	8,0 a	0,1256 a

* = Significant at the 5% probability level; ** = Significant at the 1% probability level;

NS = Not significant. Averages followed by the same letter do not differ by Tukey's test. PG(%); IVG(seeds/day); Tm(days); Vm(days).

The germination speed index (GVI) was significant at the 1% probability level and the coefficient of variation was 29.34% (Table 10). When analyzing the GVI in relation to the substrate (Figure 1) it can be seen that sand had the highest value (4.14), the composition of sand + coconut fiber (3.74), coconut fiber (3.63) and sugar cane bagasse (2.38) had the lowest performance. Helio 253 had the lowest GVI (1.41), while Olisum 3 had the highest (4.80), followed by genotypes AG 262 (4.43) and Embrapa 122-v2000 (3.26). Comparing the results with those obtained by Santos Jùnior *et al.* (2013) who studied the initial growth of sunflower (variety EMBRAPA 122-V2000) grown in a hydroponic system in various substrates and irrigated with saline water, they found the best germination speed index for the sand and sugar cane bagasse substrates. While in this experiment the best GVI were sand, coconut fiber and the sand + coconut fiber composition (Figure 4).

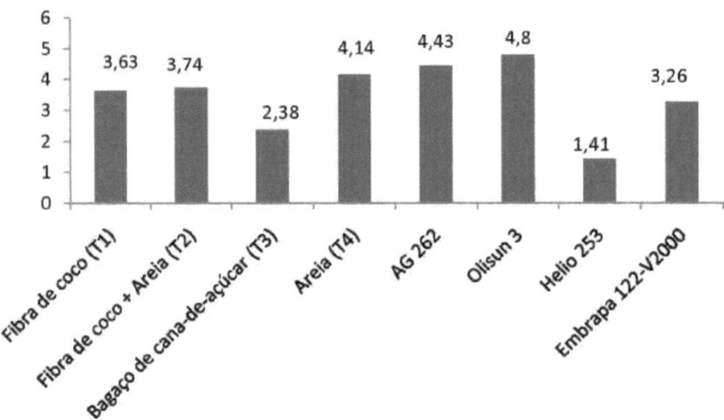

Figura 4 germination speed index in relation to substrate and sunflower genotypes.

Vieira and Krzyzanowsky (1999) state that for the IVG variable, the higher the value, the greater the ability of the seeds to express their potential. It can be inferred that sunflower seeds probably have greater ability to germinate in substrates that provide the best conditions, such as sand and coconut fiber. According to Martins *et al.* (1999), plants grown in these substrates are less vulnerable to adverse environmental conditions because they emerge more quickly and spend less time in the early stages of development.

Nogueira *et al.* (2003), evaluating different substrates for mangabeira, found that sand was responsible for the highest germination speed index. Ramos *et al.* (1983), using angico seeds (*Parpiptadenia rigida* (Benth) Brenan) at sowing depths of 0.5 and 1.5, found that the substrates sand, sawdust and burlap had a satisfactory germination speed index.

According to Table 10, the average germination time (Tm) was not statistically significant, ranging from 8.0 to 8.92, while the average speed (Vm) for the substrates was significant at 5% and for the genotypes differed statistically at 1%, remaining constant at values close to 0.1 according to Figure 5.

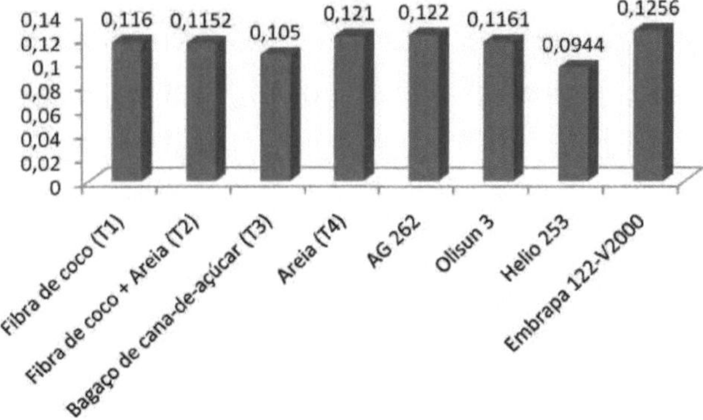

Figura 5   Average germination speed in relation to substrate and sunflower genotypes.

Figure 5 shows that the lowest average speed is obtained by the sugar cane bagasse substrate and the Helio 253 genotype. The highest average speed is obtained by the sand, coconut fiber and coconut fiber + sand substrates, as well as the Ag 262, Olisun 3 and Embrapa 122- v2000 genotypes.

Seedlings grown in pure coconut fiber substrates or in compositions with sand can make the seedlings less vulnerable to adverse environmental conditions because they emerge more quickly into the soil and spend less time in the early stages of development (MARTINS *et al.*, 1999).

CONCLUSIONS

1.   The genotypes Embrapa 122-v2000 and Helio 253 had the best germination percentage with values over 70%. The best germination speed index was for Olisum 3, AG 262 and Embrapa 122-v2000.

2.   The most suitable substrates were sand, coconut fiber and coconut fiber + sand.

3.   Sugar cane bagasse did not produce satisfactory results and could be better evaluated if used with another substrate.

4.   The use of agro-industrial waste as a substrate, such as coconut fiber and sugar cane bagasse, is a sustainable and economical alternative for small farmers.

BIBLIOGRAPHICAL REFERENCES

BACAXIXI, P.; RODRIGUES, L.R.; BUENO, C.E.M.S.; RICARDO, H. A.; EPIPHANIO, P.D.; SILVA, D.P.; BARROS, B.M.C.; SILVA, T.F. Sunflower germination test (*Helianthus annuus* L.) Revista Cientifica Eletrônica de Agronomia, n. 20, 2011.

BALBINOT JR.; BACKES, R.L.; SOUZA, A.M. Performance of sunflower cultivars in three sowing seasons in the northern plateau of Santa Catarina. Scientia Agraria, v.10, p.127-133, 2009.

BRAZIL. Ministry of Agriculture and Agrarian Reform. Rules for seed analysis. Brasilia: LAVARV/ SNAD, 1992, 365p.

CAMILLI, L.; IKEJIRI, L.; KLEIN, J.; RODRIGUES, J. D.; BOARO, C. S. F. Productivity and estimation of the in vivo carboxylation efficiency of the rubisco enzyme in ornamental sunflower grown in sewage sludge. Revista Brasileira de Biociências, v. 5, p. 858-860, 2007.

DAROS, E.; ZAMBOM, J. L. C.; KOEKLER, H. S. Response of sunflower genotypes to plant population variation. In: XI National Sunflower Research Meeting, 1995, GOIÂNIA-GO. XI National Sunflower Research Meeting - SUMMARIES. LONDRINA-PR: EMBRAPA/CNPSO, 1995. v.1. p.24.

FERREIRA, D. F. SISVAR. Version 4.3. Lavras: UFLA, 2003. Software.

LABORIAU, L. G. & VALADARES, M. B. (1976). On the germination of seeds of Calotropis procera. Annals of the Brazilian Academy of Sciences, Sao Paulo, 48:174-186.

LINHARES, P. C. F.; ABREU, W. B. de; MENEZES NETTO, A. C.; SANTOS, V. G. dos; SOUSA, A. H. de; MA\RACAJA, P. B. Substrates in the emergence and vigor of sunflower seedlings. Revista de Biologia e Ciências da Terra, Campina Grande, PB. v.5, n.1, p. 1519-1527, 2005.

MAGUIRE, J.D. (1962). Speed of germination aid in selection and evaluation for seedling emergence and vigor. Crop Science, Madison, 2(2):176-177.

MALHEIROS , S. M. P. &PAULA JÛNIOR, D. R. Utilization of the composting process with agro-industrial waste. *In* : BRAZILIAN CONGRESS OF SOIL SCIENCE, 26, 1997.

MARTINS, C. C.; NAKAGAWA, J. & BOVI, M. L. (1999). Effect of seed position on substrate and initial growth of red palm (Euterpe espiritosantensis Fernades - Palmae) seedlings. Revista Brasileira de Sementes, Brasilia, 21(1):164-173.

NOGUEIRA, R. J. M. C.; ALBUQUERQUE, M. B. & SILVA JUNIOR, J. F. (2003) Effect of substrate on emergence, growth and stomatal behavior in mangabeira seedlings. Revista Brasileira de Fruticultura, Jaboticabal, 25(1):15- 18.

RAMOS, A.; BIANCHEITTI, A.; KUNYIOSHI, Y. S. (1983). Influence of the type and thickness of bed cover on the emergence and vigor of angico seeds (Parapiptadenia rigida (Benth) Brean). In: CONGRESSO FLORESTAL

ROSSI, R. O. Sunflower. Curitiba: Tecnoagro LTDA, 1998. 333p.

SANTOS JÛNIOR, J. A.; MEDEIROS, S. S.; GUEDES FILHO, D. H.; DIAS, N. S.

Emergence of sunflower grown in a hydroponic system with various substrates under saline stress. Inovagri International Meeting. IV International Workshop on Technological Innovations in Irrigation, 2013.

SEVERINO, L. S.; LIMA, R. L. S.; BELTRÂO, N. E. M. Composiçâo Quimica de Eleze Materiais Orgânicos Utilizados em Substratos para Produçâo de Seedlings. Ministry of Agriculture, Livestock and Supply. ISSN 0102-0099. August/2006, Campina Grande, PB.

SILVA, A. R. A.; BEZERRA, F. M. L.; SOUSA, C. C. M.; PEREIRA FILHO, J. V.;FREITAS, C. A. S. Performance of sunflower cultivars under different irrigation rates in the Curu Valley, CE. Revista Ciência Agronômica, v. 42, n. 1, p. 57-64, 2011.

SILVA, R. P.; PEIXOTO, J. R.; JUNQUEIRA, N. T. V. Influence of different substrates on the development of passion fruit seedlings (*Passiflora edulis Sims* f. *flavicarpa* DEG). Revista Brasileira de Fruticultura, Jaboticabal-SP , v.23, n.2, p.377-381, 2001.

VIEIRA, R. D.; KRZYZANOWSKI, F. C. Electrical conductivity test. In: KRZYZANOWSKI,F.C.; VIEIRA, R.D.; FRANÇA NETO, J.B. (Ed.). *Seed vigor: tips and tests*. Brasilia: ABRATES, 1999. Chap. 4, p. 1-26.

CHAPTER 4

SUBSTRATE VARIETIES ON THE QUALITY OF SUNFLOWER SEEDLINGS IRRIGATED WITH WASTEWATER

ABSTRACT: The formation of quality seedlings is very important for obtaining healthy crops. The sunflower has a high capacity for climate adaptation as well as economic potential. In this context, the research was carried out in a greenhouse at the Federal University of Campina Grande, with the aim of evaluating the quality of sunflower seedlings in different substrate compositions irrigated with treated wastewater. Commercial substrate, soil and coconut fiber were used in the substrate compositions. The ratio of plant height to collar diameter (RAD), total dry mass, the aerial part/root system ratio and the Dickson quality index were evaluated. The composition of commercial substrate (S1) and commercial substrate + soil (S6) at 45 days after sowing had higher values than the others, with RAD of 8.98 and 8.66, respectively. Irrigation with wastewater led to a higher total dry mass from the third evaluation onwards. The substrates that were considered suitable for quality in the formation of sunflower seedlings were soil + coconut fiber (S4), commercial substrate + coconut fiber (5) and commercial substrate + soil (S6).

Keywords: coconut fiber, reuse, treated effluent.

INTRODUCTION

Cultivated on all continents due to its ability to adapt to different climatic conditions, the sunflower (Helianthus annuus L) originated in the United States and Mexico (ARRUDA, 2010). It has special characteristics in terms of its economic potential, as it is very dynamic in terms of the different ways it can be used, as well as being a plant well known for its ornamental potential (SILVA *et al.*, 2011).

Sunflower is capable of producing grain, vegetable oil, fodder and green manure, and can be grown at any time of the year (LOBO, 2006). However, in order to obtain high yields, it is essential to use substrates in the nursery phase with characteristics that favor seed germination, root development and initial plant growth (RAMOS *et al.*,

2002); so the best substrates must have, among other important characteristics, easy availability for purchase and transportation, absence of pathogens, richness in essential nutrients, adequate pH, good texture and structure (SILVA *et al.*, 2001).

Thus, growing plants in alternative substrates has been increasingly used in our country. The use of solid waste from agricultural and agro-industrial processes prevents the accumulation of waste, improves pollution control, improves public health conditions, reduces dependence on imported chemical fertilizers and enables sustainable development in agriculture. The use of this waste as fertilizer or substrate allows for the recovery of elements present in the waste, such as nitrogen (N), phosphorus (P), potassium (K) and trace elements. In addition, the addition of organic matter to the soil helps to improve its physical structure, water absorption capacity and the supply of nutrients to plants, making it possible to increase production and improve food quality (Malheiros and Paula Júnior, 1997).

Severino *et al.* (2006) point out that for a material to be used as a substrate for seedlings, as well as having appropriate chemical and physical characteristics, it needs to be available close to the production site in sufficient quantity, as well as being low cost.

However, it is difficult for one material alone to provide all the desirable characteristics for seedling formation and it is therefore necessary to check the best combination to be used for each species (BIASI *et al.*, 1995). In this way, mixing different components in order to obtain a suitable substrate for producing quality seedlings with adequate health in a short period of time can lead to gains in the growth and production of seedlings of this ornamental plant species and also reduce the final cost.

The use of agro-industrial waste in agricultural substrates as a source of nutrients can be used concomitantly with the planned reuse of domestic wastewater in irrigation, as it can favor water conservation and the possibility of nutrient input and recycling, contributing to the preservation of the environment (VAN DER HOEK *et al.*, 2002).

The use of water for irrigation is a practice that has been widely studied and recommended by various researchers as a viable alternative for meeting the water and, to a large extent, nutritional needs of plants (ALVES *et al.*, 2009; CAPRA and SCICOLONE, 2007; HERPIN *et al.*, 2007).

Thus, according to Fonseca (2000), obtaining quality seedlings before definitive planting can be achieved in a practical, quick and easy way just by observing morphological parameters, defining a quality seedling as one that survives and develops after planting in the field.

In this context, the aim of this study was to evaluate the quality of sunflower seedlings in different substrate compositions irrigated with treated wastewater.

MATERIALS AND METHODS

The experiment was carried out in a greenhouse belonging to the Academic Unit of Agricultural Engineering (UAEAg) of the Federal University of Campina Grande - UFCG, located in the municipality of Campina Grande, state of Paraiba - PB, at geographic coordinates 7°15'18" south latitude and 35°52'28" west longitude, at an altitude of 550 m.

The seeds of the ornamental sunflower cultivar Sol Noturno and, subsequently, the sunflower seedlings were sown and grown in tubes (black plastic material with a capacity of 285ml, duly identified), placed on a metal shelf suitable for tubes, with a height of 37.0 cm, with 252 cells, reaching the end of the experiment at the seedling stage.

The treatments corresponded to 6 combinations of substrate: 100% commercial substrate (SC) - S1, 100% soil (S) - S2, 100% coconut fiber (CF) - S3 and 50% of (S) mixed with 50% of (CF) - S4; 50% of (SC) mixed with 50% of (CF) - S5 and 50% of (SC) mixed with 50% of (S) - S6 and 2 types of water: A1 - supply water and A2 - waste water.

The commercial substrate used in the experiment was composed of simple superphosphate, potassium nitrate, peat, vermiculite and pine bark; the soil - Neossolo Regolitico Distròfico was acquired from the district of Sao José da Mata, PB and the coconut fiber, which is an agro-industrial residue, has 37% humidity, electrical conductivity equal to 0.3 mS.cm^{-1} , pH equal to 5.9 from Paraipaba - CE.

All the tubes were filled with the substrates and previously brought to field capacity with the water of the respective treatment on the day before sowing, which was carried out with 3 seeds per tube, and from then on irrigation was carried out twice a day, in the early morning and late afternoon, so that the tubes remained at field capacity, regardless of the different types of substrates. Then the seedlings were thinned out, leaving only one seedling per container - the one that was the most vigorous (tallest, greenest leaves and without signs of attack by pests and/or pathogens or the most upright) compared to the others, which could be seen with the naked eye.

Irrigation began shortly after sowing and continued until the end of the experiment, with the help of a 500 ml pissette at a rate of 15 ml per tube, every day. For this purpose, we used local water supply (A1) - from the Paraiba Water and Sewage Company (CAGEPA), located in the municipality of Campina Grande, PB - and wastewater (A2), from the Monte Santo stream, treated by the UASB anaerobic reactor, a wastewater treatment unit that is capable of removing pathogens and suspended solids, considerably reducing the polluting potential of sewage after treatment; so that they could constantly be at or near field capacity (FC).

For this purpose, soil and water analyses were carried out at the Irrigation and Salinity Laboratory belonging to the Department of Agricultural Engineering at the Federal University of Campina Grande, using the methodology presented by EMBRAPA (1997) for soil and APHA (1997) for water, according to Tables 11 and 12, respectively.

Table 11. Physical and chemical characteristics of the soil used to compose the substrate.

Soil characteristics	
Physics	
Textural classification	Clayey loam
Apparent Specific Mass - 33kPa (kg dm)^3	1,45
Porosity (%)	42,35
Field Capacity (g kg)$^{-1}$	83,6
Wilting Point (g kg)$^{-1}$	22,9

Available Water (g kg)$^{-1}$		60,7
Chemistry		
Assortative Complex (cmol$_c$ kg)$^{-1}$		
Calcium (Ca^2 +)		1,87
Magnesium (Mg)$^{2+}$		1,05
Sodium (Na)$^+$		0,06
Potassium (K)$^+$		0,23
Saturation Extract (mmol$_c$ L)$^{-1}$		
Cl$^-$		3,75
CO$_3$ $^{2-}$		Absent
HCO$_3$ -		1,70
SO$_4$ $^{2-}$		Present
Ca $^{2+}$		1,75
Mg $^{2+}$		2,00
In$^+$		1,12
K$^+$		0,55
pHps		6,15
EC $_{es}$ (dS m)$^{-1}$		0,67

Table 12. Chemical analysis of the water used in the experiment.

Month	pH	Cea	P-Total	K	N-Total	In	Ca	Mg	Zn	Cu	Fe	Mn	RAS
		(dS.m)$^{-1}$						mgL^{-1}					(mmolL)$^{-10.5}$
Water Supply													
Average	7,2	0,8	a	5,39	a	35,54	20	15,2	a	a	a	a	1,44
Treated Wastewater													
January	7,6	1,06	3,52	30,36	28,5	171,9	50,1	45,1	0,01	0,008	0,001	0,003	4,51
February	7,7	1,1	3,59	30,42	29,2	171,5	50,9	45,7	0,01	0,006	0,001	0,001	4,22
March	7,9	1,2	3,68	30,47	31,2	178,1	52,3	46,6	0,01	0,006	0,001	0,007	4,26
Average	7,73	1,12	3,60	30,42	29,63	173,83	51,10	45,80	0,01	0,007	0,001	0,004	4,33

a: absence

The experimental design adopted was a randomized block design with a factorial scheme of 6 x 2, with 3 replications, totaling 36 plots, each consisting of three tubes, for a total of 108 experimental units.

Growth was assessed at 15, 21, 30, 37 and 45 days after sowing - DAS - and the plant height/collector diameter ratio (RAD), obtained from the ratio between the height of the aerial part and the diameter of the collar.

Three evaluations (15, 30 and 45 DAS) were carried out to assess the quality of the seedlings in relation to the fresh and dry phytomass of the aerial part and root system, determining the following characteristics: the total dry mass (MST), obtained from the sum of the dry mass of the aerial part and the dry mass of the root system; the aerial part/root system ratio (RPAR), obtained from the ratio between the dry mass of the aerial part and the dry mass of the root system; and the Dickson quality index (IQD), according to Dickson *et al.*, (1960), where DQI = MST/(RAD+RPAR). The RAD used in the DQI was from the last growth assessment.

The height of the seedlings was determined from the neck of the plants to the meristematic apex and the cotyledonary leaves were excluded. The stem of each plant was cut off close to the ground, and the fresh weight of the aerial part and the root were weighed immediately using a 0.01 g precision analytical balance. After weighing the fresh phytomass, the plants were placed in duly identified paper bags and dried in a forced-air oven at 65 °C until a constant weight was obtained. The dry phytomass of the plants was then determined using an analytical balance with an accuracy of 0.0001g.

The results were evaluated by analysis of variance and the means were compared using the Tukey test at 5% significance, with the aid of the SISVAR statistical software (FERREIRA, 2003).

RESULTS AND DISCUSSION

Table 13 shows the summary of the analysis of variance for the plant height/collar diameter ratio (RAD) in the five evaluation periods, with a significant effect ($p < 0.05$) in the first and fourth evaluations for the variation factor Substrate, so that the commercial substrate (S1), followed by the commercial substrate + soil composition (S6), were the ones that stood out the most, reaching 45 days after sowing - DAS, with a RAD of 8.98 and 8.66, respectively.

Table 13. Summary of the analysis of variance for the ratio of plant height to crown diameter (RAD) in the five evaluation periods for sunflower seedlings grown in different substrates.

Mean Squares

Source of Variation	GL	RADI	RAD2	RAD3	RAD4	RAD5
Water (A)	1	0,381 ns	0,080ns	0,703ns	10,604ns	22,154*
Substrates (S)	5	1,578*	2,879ns	3,154ns	7,204*	11,170ns
Water x Substrates	5	0,426ns	0,794ns	1,513ns	2,869ns	3,158ns
Waste	24	0,519	1,130	1,804	2,579	4,900
CV (%)		25,72	26,73	27,71	26,30	29,20

Type of substrate				Average		
Commercial substrate (S1)		3,62 a	5,11a	5,64 a	7,34 a	8,98 a
Soil (S2)		2.56 ab	3,30 a	3,64 a	4,33 b	5,40 a
Coconut fiber (S3)		2,13 b	3,50 a	5,06 a	5.71 ab	7,74 a
Soil + coconut fiber (S4)		2.81 ab	3,83 a	4,98 a	6.64 ab	8,17 a
Commercial substrate + coconut fiber (S5)		2.56 ab	3,62 a	4,38 a	5.56 ab	6,52 a
Commercial Substance + Soil (S6)		3.09 ab	4,48 a	5,36 a	6.94 ab	8,66 a

(**) significant at 1%, (*) significant at 5% and (ns) not significant by the F test; CV - coefficient of variation;

For the Water factor, there was only a statistical difference at 45 days after sowing (Table 13). However, Figure 6 shows that the seedlings irrigated with wastewater had superior RAD from 30 DAS onwards, probably due to the presence of nutrients in the water, which induces greater plant growth.

According to the standard recommended by Birchler *et al.* (1998), the ratio of plant height to collar diameter (RAD) should be less than 10, which agrees with the results found in this research for all the substrates studied, as well as for the types of water. Carneiro (1995) comments that for a good result for this ratio, the average values should be between 5.4 and 8.1; Therefore, wastewater can be used to irrigate seedlings, as it avoids or reduces the use of potable water, reduces its disposal in water bodies, thus reducing pollution and also reduces spending on fertilizers, due to the existence of microorganisms and nutrients in the water itself.

The sunflower seedlings grown in different substrates also obtained an optimum RAD, from the point of view of the standard recommended by Birchler et al. (1998), for all the evaluation dates, but within the 5.4 to 8.1 range, only those grown in soil, coconut fiber and commercial substrate + coconut fiber fit in, with RAD of 5.4, 7.74 and 6.52, respectively, at 45 DAS (Table 13).

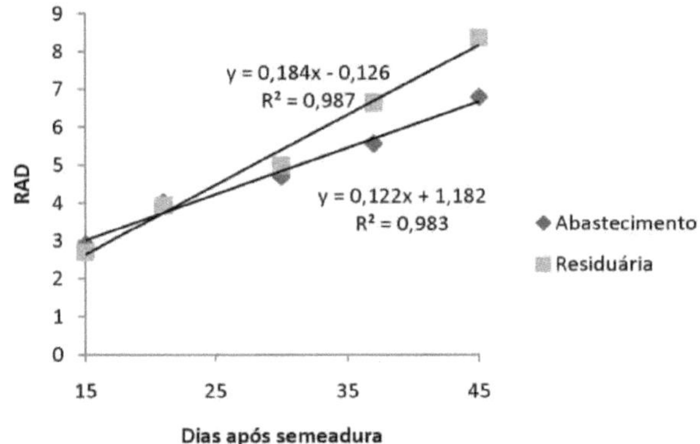

Figura 6.   Plant height/collector diameter ratio for sunflower seedlings irrigated with wastewater and supply water.

The value resulting from dividing the height of the plant by its respective collar diameter expresses the balance of growth, relating these two important morphological parameters in just one index (CARNEIRO, 1995), also known as the robustness quotient, which is considered one of the most accurate, as it provides information on how thin the seedling is (JOHNSON & CLINE, 1991).

Table 14 shows the summary of the analysis of variance for total dry mass (MST) and the ratio of the weight of dry mass of the aerial part to the weight of dry mass of the root system (RPAR) for sunflower seedlings on the three evaluation dates, showing that for the MST variable, there was a significant effect at the 1% probability level only for the factor Substrates (S); S6 - commercial substrate + soil - stood out in all evaluations, with the highest average MST. There was a greater disparity in this variable at 45 DAS, between S6 and S5 (commercial substrate + coconut fiber), when S5 reached only 10.68% of the MST value of S6.

Irrigation with wastewater led to greater total dry mass only on the third and last evaluation dates, but this was not statistically significant (Table 14).

With regard to the weight of the dry mass of the aerial part and the weight of the dry mass of the root system (RPAR), there was a significant effect for the Water factor ($p<0.05$), at 15 DAS, with a higher average for the seedlings irrigated with wastewater, however, this superiority was not due to this factor, as it occurred at a very early stage of growth and was not repeated in the following evaluations. The weight of the dry mass is a good indication of the resistance capacity of seedlings in field conditions, even if it is a destructive method (GOMES, 2001).

The RPAR for the Substrate Type (S) factor showed a statistical difference ($p<0.01$) at 15 and 30 DAS (Table 14). According to Brissette (1984), for the ratio of the weight of the dry mass of the aerial part to the weight of the dry mass of the root (RPAR), the value that best represents this ratio is equal to 2.0. This ratio is considered to be an efficient and reliable index for expressing the quality standard of these seedlings (PARVIAINEN, 1981). In comparison with the authors, all the results obtained (Table 13) in relation to the substrates ranged from 1.66 to 3.47 at the end of the experiment, verifying that the seedlings grown in the combination of commercial substrate + coconut fiber (S5) were very close to this value - 1.98, so they are at an efficient and safe standard.

The soil + coconut fiber substrate (S4) had a high RPAR value on all evaluation dates, showing that this substrate is not suitable for sunflower seedlings.

Table 14. Summary of the analysis of variance for the total dry mass (MST), the ratio of the weight of the dry mass of the aerial part to the weight of the dry mass of the root system (RPAR) for sunflower seedlings grown in different substrates.

		Mean Squares					
Source of Variation	GL	MST^1_1	MST12	MST^2_3	RPAR21	$RPARi_2$	RPAR23
Water (A)	1	$0,0004^{ns}$	$0,0071^{ns}$	$0,0117^{ns}$	$0,1690^*$	$0,0182^{ns}$	$0,0014^{ns}$
Substrates (S)	5	$0,0151^{**}$	$0,1821^{**}$	$0,1144^{**}$	$1,8990^{**}$	$0,466^{**}$	$0,2841^{ns}$
Water x Substrates	5	$0,0067^{ns}$	$0,0053^{ns}$	$0,0154^{ns}$	$0,2595^{**}$	$0,0401^{ns}$	$0,1480^{ns}$

| Waste | 24 | 0,0026 | 0,0045 | 0,0158 | 0,0686 | 0,0650 | 0,1087 |
| CV (%) | | 20,48 | 16,56 | 13,81 | 19,41 | 21,05 | 19,85 |

Type of water				Average			
Water Supply - A1		0,070 a	0,211 a	0,324a	1,30b	1,61 a	2,47a
Residual Water - A2		0,065 a	0,176 a	0,386 a	2,04 a	1,55 a	2,31 a

Type of Substrate				Average			
Commercial substrate (S1)		0,063b	0,098 c	0,155 c	0,72b	1,11 c	2,63 a
Soil (S2)		0,044 b	0,168 c	0.262 bc	1,11b	1.33 bc	1,66 a
Coconut fiber (S3)		0,053 b	0,061 c	0.263 bc	0,51b	1.23 bc	1,71 a
Soil + coconut fiber (S4)		0,071 b	0,300 b	0.643 ab	5,94 a	2.32 ab	3,47 a
Composite Substrate + Coconut Fiber (S5)		0,053 b	0,066 c	0,078 c	0,62 b	0,86 c	1,98 a
Commercial Substance + Soil (S6)		0,120 a	0,474 a	0,730 a	1,11 b	2,61 a	2,89 a

Variables transformed into the root of x;[2] Variables transformed into the root of x + 0.5;[3] Variables transformed into the root of x + 1.0; GL - degree of freedom; significant at 1% (**), at 5% (*) and (ns) not significant by the F test; CV - coefficient of variation;

According to Maeda *et al*, (2006) the aggregation of the root system to the substrate is essential to guarantee the survival and development of the seedlings after planting in the field, so it is essential that the seedlings have a good RPAR value, as this parameter is directly related to quality, in terms of root weight, It should be noted that when the seedlings go to the field, the aerial part should not be too much larger than the root, otherwise there may be problems with water absorption in the aerial part.

There was no statistical significance for the Water factor, however, for the Substrate factor there was a significant difference at 1% probability level, with IQD varying from 0.010 to 0.063 (Table 15).

Table 15. Summary of the Dickson Quality Index (DQI) analyses for sunflower seedlings grown in different substrates.

Mean Squares

Source of Variation	GL	IQD 1[1]	IQD½	IQD 3[2]
Water (A)	1	0.000006ns	0.001246ns	0.000038ns
Substrates (S)	5	0,003819**	0,020144**	0,001337**
Water x Substrates	5	0,003111*	0,001643*	0.000214ns

| Waste | 24 | 0,00091 | 0,00060 | 0,00026 |
| CV (%) | | 19,87 | 15,27 | 2,19 |

Type of water			Average	
Water Supply - A1		0,025a	0,032a	0,032a
Wastewater - A2		0,024a	0,026a	0,035a

Type of Substrate			Average	
Commercial substrate (S1)		0,018b	0,014c	0.013bc
Soil (S2)		0,017b	0,033b	0.033abc
Coconut fiber (S3)		0.024ab	0,010c	0.026abc
Soil + coconut fiber (S4)		0.026ab	0,041b	0.056bc
Composite Substrate + Coconut Fiber (S5)		0,021b	0,014c	0,010c
Commercial Substance + Soil (S6)		0,040a	0,062a	0,063a

[1] Variables transformed into root of x;[2] Variables transformed into root of x + 0.5; GL - degree of freedom; significant at 1% (**), at 5% (*) and (ns) not significant by F test; CV - coefficient of variation;

The Dickson Quality Index is mentioned as a promising integrated morphological measure (JOHNSON & CLINE, 1991) and is considered a good indicator of seedling quality, as its calculation takes into account the robustness and balance of the biomass distribution in the seedling, weighing up the results of several important parameters used to assess quality (FONSECA et. al., 2002). According to Gomes (2001), the higher the Dickson quality index, the better the quality of the seedlings, so the best values were obtained for Soil + Coconut Fiber (S4) and Commercial Substrate + Soil (S6).

CONCLUSION

1.   In terms of plant height and collar diameter, all the substrates evaluated had satisfactory results.

2.   For the ratio of the weight of the dry mass of the aerial part to the weight of the dry mass of the root, the commercial substrate + coconut fiber (S5) had an average of 1.98, considered ideal for producing sunflower seedlings.

3.   The highest Dickson Quality Index scores were obtained with the substrates soil + coconut fiber (S4) and commercial substrate + soil (S6), with averages of 0.056 and

0.063, respectively, making them the best substrates for obtaining quality seedlings.

4.   The most suitable substrates for the quality of sunflower seedling formation were soil + coconut fiber (S4), commercial substrate + coconut fiber (5) and commercial substrate + soil (S6).

5.   The use of wastewater influences the availability of nutrients for the plant, as well as enabling the production of seedlings, reducing environmental impacts and saving drinking water.

BIBLIOGRAPHICAL REFERENCES

ALVES, W. W. A. et al. Leaf area of cotton irrigated with wastewater fertilized with nitrogen and phosphorus. Revista Verde de Agroecologia e Desenvolvimento Sustentàvel, Mossoró, v. 4, n. 1, p. 41-46, 2009.

APHA - American Public Health Association. Standard methods for the examination of water and wastewater. 20 ed. New York. APHA, AWWA, WPCR, 1997, 1994p.

ARRUDA, N.T.; OLIVEIRA, F.A.; BATISTA, J. A.; MENEZES, Ê.F.; RODRIGUES, A.F. Growth rate and nutrient content in the plant of sunflower crop submitted to applications of limestone and phosphorus. Revista Verde de Agroecologia e Desenvolvimento Sustentàvel ISSN 1981-8203 (Mossoró - RN - Brasil) v.5, n.4, p. 179 -184 October/December 2010.

BIASI, L. A. et al. Effect of mixtures of peat and sugarcane bagasse on the production of passion fruit and tomato seedlings. Scientia Agricola, Piracicaba, v. 2, n. 52, p. 239-243,1995.

BIRCHLER, T.; ROSE, R. W.; ROYO, A.; PARDOS, M. La planta ideal: revision Del concepto, parâmetros definitorios e implementaction practica. Investigacion Agraria, Sistemas y Recursos Forestales, Madrid, v. 7, n. 1/2, p. 109121, 1998.

BRISSETTE, J. C. Summary of discussions about seedling quality. In: SOUTHERN NURSERY CONFERENCES, 1984, Alexandria. Proceedings...New Orleans: USDA Forest Service, Southern Forest Experiment Station, 1984. p.127128.

CAPRA, A.; SCICOLONE, B. Recycling of poor quality urban wastewater by drip irrigation systems. Journal of Cleaner Production, v. 15, n. 16, p. 1529-1534, 2007.

CARNEIRO, J. G. A. Production and quality control of forest seedlings. Curitiba: UFPR/FUPEF, 1995. 451 p.

DICKSON, A.; LEAF, A.L.; HOSNER, J. F. Quality appraisal of white spruce and white pine seedling stock in nurseries. Forestry Chronicle, v. 36, p. 10-13, 1960.

EMBRAPA - Brazilian Agricultural Research Corporation. Manual and methods of soil analysis. 2 ed. Rio de Janeiro. National Soil Research Center. 1997, 247p.

FERREIRA, D.F. Programa Sisvar - statistical analysis program. Lavras: UFLA, 2003.

FONSECA, E.P. Quality standards for seedlings of Trema micrantha (L.) Blume, Cedrela fissilis Vell. and Aspidosperma polyneuron Müll. Arg. produced under different shading periods.

FONSECA, E. P.; VALÉRI, S. V.; MIGLIORANZA, E.; FONSECA, N. A. N.; COUTO, L. Quality standard of Trema micrantha (L.) Blume seedlings, produced under different shading periods. *Revista Arvore*, 26,515-523. 2002.

GOMES, J. M. Parâmetros morfológicos na avaliaçao da qualidade de mudas de Eucalyptus grandis, produzidas em diferentes tamanhos de tubete e de dosagens de N-P-K. 2001.166f. Thesis (Doctorate in Forest Science) - Federal University of Viçosa, Viçosa, 2001.

HERPIN, V. et al. Chemical effects on the soil-plant system in a secondary treated wastewater irrigated coffee plantation - a pilot field study in Brazil. Agricultural Water Management, Amsterdam, v. 89, n.1, p. 105-115, 2007.

HUNT, G.A. Effect of styroblock design and cooper treatment on morphology of conifer seedlings. In: TARGET SEEDLING SYMPOSIUM, MEETING OF THE WESTERN FOREST NURSERY ASSOCIATIONS, GENERAL TECHNICHAL REPORT RM-200, 1990, Roseburg. Proceedings... Fort Collins: United States Department of Agriculture, Forest Service, 1990. p.218-222.

JOHNSON, J. D.; CLINE, P. M. Seedling quality of southern pines. In: DUREYA, M. L., DOUGHERTY, P. M. (Eds.). Forest regeneration manual. Dordrecht: Kluwer Academic Publishers, 1991. p. 143-162.

LOBO, T. F. Levels of sewage sludge in the development, nutrition and productivity of the sunflower crop. 2006. 64 f. Dissertation (Master's Degree in Agronomy) - Faculty of Agronomic Sciences, São Paulo State University, Botucatu, 2006.

MAEDA, S.; ANDRADE, G.C.; FERREIRA, C.A.; SILVA, H. D.; AGOSTINI, R.B. Alternative substrates for the production of Eucalyptus badjensis seedlings, obtained from waste from the wood and brewing industries and from goat farming. Colombo: Embrapa Florestas, 2006. 5p. Technical communication, 157.

MALHEIROS , S. M. P. and PAULA JÛNIOR, D. R. Use of the composting process with agro-industrial waste. *In* : BRAZILIAN CONGRESS OF SOIL SCIENCE, 26, 1997.

PARVIAINEN, J. V. Quality and quality evaluation of forest seedlings. In: SEMINARIO DE SEMENTES E VIVEIROS FLORESTAIS, 1, 1981, Curitiba. Anais...Curitiba: FUPEF, 1981. p.59-90.

RAMOS, J.D. et al. Seedling production of fruit plants. Informe Agropecuàrio, Belo Horizonte, v. 23, p. 64-72, 2002.

SEVERINO L. S; LIMA, R. L.; BELTRÂO N. E. M. Composiçao Quimica de Eleze Materiais Orgânicos Utilizados em Substratos para Produçao de Mudas. Technical communication 27, EMBRAPA, Campina Grande-PB, 2006.

SILVA, R. P. et al. Influence of different substrates on the development of passion fruit seedlings (Passiflora edulis Sims f. flavicarpa DEG). Revista Brasileira de Fruticultura, Jaboticabal, v.23, n.2, p.377- 381, agosto 2001.

SILVA, A. R. A.; BEZERRA, F. M. L.; SOUSA, C. C. M.; PEREIRA FILHO, J. V.;FREITAS, C. A. S. Performance of sunflower cultivars under different irrigation rates in the Curu Valley, CE. Revista Ciência Agronòmica, v. 42, n. 1, p. 57-64, 2011.

VAN DER HOEK., W. et al. Urban wastewater: a valuable resource for agriculture; a casestudy from horoonabad, Pakistan. Colombo: International Water Management Institute. 2002. 29 p. (Research Report, 63).

CHAPTER 5

PHYTOMASS PRODUCTION OF ORNAMENTAL SUNFLOWER SEEDLINGS GROWN IN DIFFERENT SUBSTRATES UNDER WASTEWATER IRRIGATION

ABSTRACT: The aim of this study was to evaluate the effect of using wastewater and its interaction with different types of substrate on the phytomass of ornamental sunflower seedlings. In a greenhouse belonging to the Agricultural Engineering Academic Unit (UAEAg) of the Federal University of Campina Grande - UFCG-PB, the following treatments were studied: 100% commercial substrate (SC), 100% soil (S), 100% coconut fiber (FC) and 50% (SC) mixed with 50% (FC), combined with 2 types of water (A1 - supply; A2 - wastewater). The experimental design was in randomized blocks, in a 4 x 2 factorial scheme, with 3 replications and 3 plants per replication. The variables analyzed were root length (RC), fresh phytomass of the aerial part (FFPA) and root (FFR), dry phytomass of the aerial part (FSPA) and root (FSR). The use of wastewater, although not significantly, resulted in higher CR averages compared to the use of supply water. The fresh and dry phytomass of the aerial part and root did not differ significantly in terms of the type of water source throughout the experiment, unlike the results obtained using the various substrates, which showed significance for some variables in at least one evaluation period.

Keywordsfomass, sunflower, substrate, wastewater

INTRODUCTION

In the state of Paraiba, the cultivation of ornamental plants has been growing in some municipalities, expanding the activity and the economy of the region. The sunflower (*Helianthusannuus L.)* is an annual plant and its cultivation, as well as that of other flowers and ornamental plants, has been widespread throughout the country due to the diversity of the Brazilian climate.

The use of wastewater in the cultivation of sunflowers, as well as various flowers and ornamental plants, is a measure that seeks to make use of wastewater for agricultural purposes, which makes it possible to provide and recycle nutrients and conserve available water of good quality.

The commercial production of seedlings and soilless cultivation of vegetables are becoming common practices among growers (CARRIJO *et al.,* 2002). The use of industrial agricultural waste as a substrate is also an alternative for recycling, as well as making the production of ornamental plant seedlings viable.

According to Rosa *et al.* (2002), recently, a substrate based on dried coconut shells from the agricultural industry has been introduced, which is rich in nutritional terms and is capable of minimizing the environmental impact caused by the solid agro-industrial waste generated. Currently, the residue or powder of mature coconut shells has been recommended as an agricultural substrate, mainly because it has an advantageous physical structure, providing high porosity, high moisture retention potential and because it is biodegradable.

The aim of this study was to evaluate the effect of using wastewater and its interaction with different types of substrate on the phytomass of ornamental sunflower seedlings.

MATERIAL AND METHODS

The experiment was conducted in a greenhouse belonging to the Agricultural Engineering Academic Unit (UAEAg) of the Federal University of Campina Grande - UFCG, located in the state of Paraiba-PB. The following treatments were studied: 100% commercial substrate (SC) - S1, 100% soil (S) - S2, 100% coconut fiber (CF) - S3 and 50% (SC) mixed with 50% (CF) - S4, combined with 2 types of water, A1 - water supply and A2 - treated wastewater.

The wastewater came from the Bodocongó reservoir in Campina Grande, PB. It was considered wastewater from the city's sewage system and was treated by the WASB to be applied to irrigate the crops.

After the containers had been filled with the respective substrate compositions and brought to field capacity, the seeds were sown the following day, with 3 seeds per tube. At 7 days after sowing (DAS), thinning was carried out, leaving only the plant that visually showed the best characteristics.

The experimental design adopted was a randomized block design with a 4 x 2 factorial scheme, with 3 replications and 3 plants per replication.

Phytomass evaluations were carried out at 15 DAS, 30 DAS and 45 DAS, i.e. at three different times during the crop's development, assessing root length (RC), fresh phytomass of the aerial part (FFPA) and root (FFR), dry phytomass of the aerial part (FSPA) and root (FSR). The stem of each plant was cut off close to the ground using a stylus, and the fresh weight of the aerial part and the root was taken immediately using an analytical balance with a precision of 0.01g. After weighing the fresh phytomass, the plants were placed in duly identified paper bags and dried in a forced-air oven at 65 °C until a constant weight was obtained, and the dry phytomass of the plants was determined using an analytical balance accurate to 0.0001g.

The results were evaluated by analysis of variance and the means were compared using the Tukey test at 5% significance, with the help of the SISVAR statistical software.

RESULTS AND DISCUSSION

Table 16 shows the summary of the analysis of variance for the three evaluation periods for the root length (RC) variable. For the water type factor, there was no significant effect on any evaluation date. For the Substrate Type factor, there was only a significant effect at 15 DAS, and there was no interaction between the variation factors studied.

Table 16. Summary of ANAVA for root length (RL), in 3 evaluation periods, of sunflower seedlings grown in four different types of substrate and irrigated with two types of water.

Variation Factor	GL	Mean Squares		
		CR_i^1	CR_2^1	$CR3^1$
Water Type (A)	1	0.00046ns	0.68ns	4.10ns
Type of Substrate (S)	3	2,48*	0.64ns	1.13ns
A x S interaction	3	0.42224ns	0.21ns	0.61ns

| Waste | 16 | 0,51 | 0,71 | 1,25 |
| C.V. | | 15,24 | 18,11 | 23,24 |

1 Variables with x-root transformation; NS: not significant (P>0.05); *: significant (P<0.05); **: significant (P<0.01); C.V.: coefficient of variation.

Wendt *et al.* (2005) obtained significant effects on the length of sunflower roots grown under green manure with oats and spontaneous vegetation. Agricultural residues, as Bayer and Mielniczuk (1999) state, provide better conditions for humidity, temperature, infiltration, soil structure and soil biological activity.

Figure 7 shows that the CR averages, in the seasons evaluated, fluctuated according to the substrate used, where S4 (SC + FC) had the highest average, not differing from the use of coconut fiber alone, with averages above 25 cm.

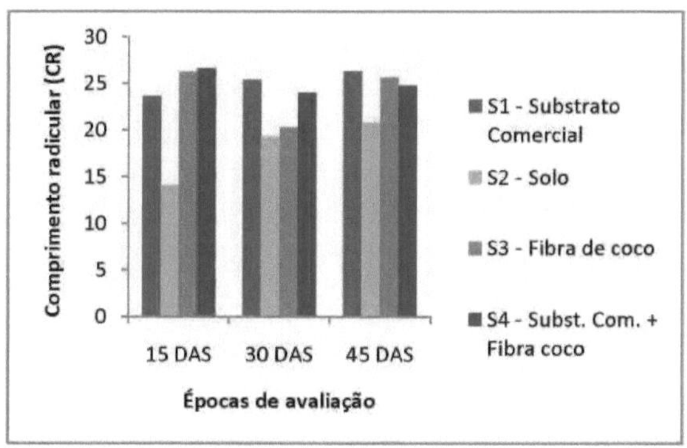

Figura 7.   Averages of the CR variable, for the Substrate Type factor in the 3 evaluation periods

Table 17 shows the summary of the analysis of variance for the variables fresh phytomass of the aerial part (FFPA) and fresh phytomass of the root (FFR), showing a non-significant interference from the Water Type factor, similar to what happened with CR. Although it had no influence on these phytomass parameters, the use of wastewater to irrigate sunflower seedlings is an environmentally restorative action that can prevent it from being disposed of excessively in water bodies. In addition to this advantage, Azevedo and Oliveira (2005) also highlight the importance of using domestic wastewater to supply nutrients and increase plant productivity.

For the Substrate Type factor, there was a significant difference for FFPA only at 30 DAS, with S2 standing out as having the highest average weight value, which did not differ from the commercial substrate.

In research carried out by Andrade *et al.* (2012) on sunflower cultivation irrigated with wastewater, they noted that there was an accumulation of dry matter in all the varieties of ornamental sunflower studied, even though there was no statistically significant effect.

Table 17. ANAVA summary for the fresh phytomass of the aerial part (FFPA) and the fresh phytomass of the root (FFR), in 3 evaluation periods, of sunflower seedlings grown in different substrates and irrigated with two qualities of water.

		Mean Squares					
Variation Factor	GL	FFPAi[1]	FFPA2[1]	FFPA3[3]	FFR1[2]	FFR2[2]	FFR3[3]
Water Type (A)	1	0.01631ns	0.01786ns	0.01047ns	0.00412ns	0.03779ns	0.04030ns
Type of Substrate (S)	3	0.02480ns	0,18685**	0.08700ns	0.08558ns	0.07982ns	0.02567ns
A x S interaction	3	0.02814ns	0.01904ns	0.12796ns	0.01612ns	0.01330ns	0.16261ns
Waste	16	0,0121	0,0219	0,1089	0,0290	0,0370	0,1109
C.V.		22,74	19,35	23,02	17,71	16,30	24,42
Type of water		Averages (unit)					
Supply		0,280a	0,672a	1,268a	0,442a	0,838a	1,152a
Residuary		0,220a	0,581a	1,044a	0,476a	1,026a	0,765a
Type of Substrate		Averages (unit)					
Commercial substrate (S1)		0,332a	0.743ab	0,918a	0,488a	1,172a	0,658a
Soil (S2)		0,252a	0,960b	1,635a	0,142a	1,172a	0,868a
Coconut fiber (S3)		0,160a	0,443a	1,282a	0,718a	0,752a	1,208a
Subst.com.+ Coconut fiber (S4)		0,257a	0,355a	0,788a	0,487a	0,632a	1,100a

[1]Variables transformed into the root of x; [2]Variables transformed into the root of x +0.5; [3]Variables transformed into the root of x +1NS: not significant (P>0.05); *: significant (P<0.05); **: significant (P<0.01); C.V.: coefficient of variation.

The quantification of dry matter is important information because sunflower can be used for green manure and silage production. As can be seen in Table 18, just like fresh phytomass, dry phytomass was not affected by the Type of Water factor, which differs from the results found by Andrade *et al.* (2007), in their work on the germination and growth of sunflower irrigated with wastewater, in which significant effects were observed in terms of fresh and dry root phytomass, compared to plants irrigated with supply water. Although the wastewater used for irrigation did not affect the vegetative variables, this could make it possible to replace fresh water and save on fertilizers.

Table 18. ANAVA summary for aerial part dry phytomass (FSPA) and root dry phytomass (FSR) in 3 evaluation periods, of sunflower seedlings grown in different substrates and irrigated with two water qualities.

		Mean Squares					
Variation Factor	GL	$FSPA_1$[1]	$FSPA_2$[1]	$FSPA3$[2]	FSR_1[1]	FSR_2[1]	$FSR3$[2]
Water Type (A)	1	0.00079ns	0.00288ns	0.00026ns	0.00004ns	0.00018ns	0.00014ns
Substrate type (S)	3	0.00199ns	0.02048**	0.00731ns	0.00223ns	0.01160**	0.00228ns
A x S interaction	3	0.00243ns	0.00210ns	0.00822ns	0.00679ns	0.00037ns	0.00229ns
Waste	16	0.0007	0.0031	0.0070	0.0038	0.0012	0.0012
C.V.		21,25	25,57	10,67	34,59	16,22	4,74
Type of water		Averages (unit)					
Supply		0,0199a	0,0578a	0,1350a	0,037a	0,044a	0,061a
Residue		0,0167a	0,0461a	0,1160a	0,033a	0,048a	0,067a
Type of Substrate		Averages (unit)					
Commercial substrate (S1)		0,0242a	0.0521ab	0,104a	0,039a	0,046a	0,051a
Soil (S2)		0,0181a	0,0943b	0,170a	0,026a	0,078b	0,093a
Coconut fiber (S3)		0,0114a	0,0308a	0,178a	0,042a	0,031a	0,085a
Subst.com.+ Coconut fiber (S4)		0,0193a	0,0305a	0,050a	0,034a	0,035a	0,029a

[1] Variables transformed into root of x; [2] Variables transformed into root of x +0.5; N^S : not significant (P>0.05);

** significant (P<0.05); : significant (P<0.01); C.V.: coefficient of variation.

The dry phytomass of the aerial part and root was influenced by the Substrate Type factor (p<0.01), only on the 2nd[a] evaluation date, with soil standing out with the best averages, thus, the non-significance in most of the evaluation periods showed the similarity of the weight averages resulting from the use of the different substrates, with practically none of them standing out. There was no significant effect of the interaction between the factors type of water and type of substrate on any of the phytomass variables studied, which indicates the independence of the factors.

Different results were obtained by Souza *et al.* (2010) when they studied the BRS OASIS sunflower with significantly better results for the use of wastewater compared to supply water in the variables analyzed at two stages of crop development.

CONCLUSIONS

1. The use of wastewater did not significantly affect any of the variables studied (CR, FFPA, FFR, FSPA, FSR), however, its use can replace fresh drinking water in the irrigation of sunflowers, as well as contributing to the environment by avoiding its incorrect disposal.

2. It was not possible to indicate the best or worst type of substrate, since they showed similar means for all the variables.

BIBLIOGRAPHICAL REFERENCES

ANDRADE, L.O.; GHEYI, H.R.; NOBRE, R.G.; DIAS, N.S.; NASCIMENTO, E.C.S. Flower quality of ornamental sunflowers irrigated with wastewater and water supply. IDESIA, v.30, n.2, p.19-27, 2012.

ANDRADE, L.; NOBRE, R.G.; SOARES, F.A.L.; GHEYI, H.R.; FIGUEIREDO, G.R.G.; SILVA, L.A. Germination and initial growth of sunflower plants (*Helianthusannuus* L.) irrigated with wastewater. Educaçao Agricola Superior, v.22, n.02, p.48-50, 2007.

AZEVEDO, L. P. de; OLIVEIRA, E. L. de. Effects of the application of sewage treatment effluent on soil fertility and cucumber productivity under subsurface irrigation. Engenharia Agricola, v. 25, n. 01, p. 253-263, 2005.

BAYER, C.; MIELNICZUK, J. Dynamics and function of organic matter. In: SANTOS, G.A.; AMARGO, F.A.O. (Ed.). Fundamentals of soil organic matter.

"Tropical and subtropical ecosystems". Ed. Porto Alegre: Gênesis, 1999.

CARRIJO OA; LIZ RS; MAKISHIMA N. 2002. Green coconut shell fiber as an agricultural substrate. Horticultura Brasileira 20: 533-535.

ROSA, M.F.; BEZERRA, F.C.; CORREIA, D.; SANTOS, F.J.S.; ABREU, F.A.P.; FURTADO, A.A.L.; BRiGIDO, A.K.L.; NORÔES, E.R.V. Utilizaçao da casca de coco como substrato agricola. Embrapa Agroindustria Tropical. Documentos, 52. Fortaleza, 24 p., 2002.

SOUZA, R.M.; NOBRE, R.G.; GHEYI, H.R.; DIAS, N.S.; SOARES, F.A.L. Utilization of wastewater and organic fertilizer in sunflower cultivation. **Revista Caatinga**, Mossoró, v.23, n.2, p.125-133, 2010.

WENDT, V.; *et al.*Sunflower yield in two sowing systems as a function of winter green manure associated with doses of NPK. **Acta Sci. Agronomia**, v.27, n.4, p.617-621, 2005.

yes
I want morebooks!

Buy your books fast and straightforward online - at one of world's fastest growing online book stores! Environmentally sound due to Print-on-Demand technologies.

Buy your books online at
www.morebooks.shop

Kaufen Sie Ihre Bücher schnell und unkompliziert online – auf einer der am schnellsten wachsenden Buchhandelsplattformen weltweit! Dank Print-On-Demand umwelt- und ressourcenschonend produzi ert.

Bücher schneller online kaufen
www.morebooks.shop

Printed by Books on Demand GmbH, Norderstedt / Germany